# Sustainable Civil Infrastructures

**Sustainable Civil Infrastructures (SUCI)** is a series of peer-reviewed books and proceedings based on the best studies on emerging research from all fields related to sustainable infrastructures and aiming at improving our well-being and day-to-day lives. The infrastructures we are building today will shape our lives tomorrow. The complex and diverse nature of the impacts due to weather extremes on transportation and civil infrastructures can be seen in our roadways, bridges, and buildings. Extreme summer temperatures, droughts, flash floods, and rising numbers of freeze-thaw cycles pose challenges for civil infrastructure and can endanger public safety. We constantly hear how civil infrastructures need constant attention, preservation, and upgrading. Such improvements and developments would obviously benefit from our desired book series that provide sustainable engineering materials and designs. The economic impact is huge and much research has been conducted worldwide. The future holds many opportunities, not only for researchers in a given country, but also for the worldwide field engineers who apply and implement these technologies. We believe that no approach can succeed if it does not unite the efforts of various engineering disciplines from all over the world under one umbrella to offer a beacon of modern solutions to the global infrastructure. Experts from the various engineering disciplines around the globe will participate in this series, including: Geotechnical, Geological, Geoscience, Petroleum, Structural, Transportation, Bridge, Infrastructure, Energy, Architectural, Chemical and Materials, and other related Engineering disciplines.

**SUCI series is now indexed in SCOPUS and EI Compendex.**

More information about this series at http://www.springer.com/series/15140

Hadi Khabbaz · Yang Xiao ·
Jia-Ruey Chang
Editors

# Smart and Green Solutions for Civil Infrastructures Incorporating Geological and Geotechnical Aspects

## Proceedings of the 6th GeoChina International Conference on Civil & Transportation Infrastructures: From Engineering to Smart & Green Life Cycle Solutions – Nanchang, China, 2021

Springer

*Editors*
Hadi Khabbaz
School of Civil Engineering
University of Technology Sydney (UTS)
Sydney, NSW, Australia

Yang Xiao
School of Civil Engineering
Chongqing University
Chongqing, China

Jia-Ruey Chang
Graduate Institute of Architecture
National Ilan University
Yilan, Taiwan

ISSN 2366-3405 ISSN 2366-3413 (electronic)
Sustainable Civil Infrastructures
ISBN 978-3-030-79649-5 ISBN 978-3-030-79650-1 (eBook)
https://doi.org/10.1007/978-3-030-79650-1



This Springer imprint is published by the registered company Springer Nature Switzerland AG
The registered company address is: Gewerbestrasse 11, 6330 Cham, Switzerland

# Introduction

Advancement in design and construction to embrace the impact of rapid global urbanisation growth in infrastructure development is inevitable. The proceedings include many smart and green solutions for civil infrastructures, incorporating geotechnical and engineering geology aspects. The articles presented in this volume are the attempts made by the researchers and practitioners to address many geotechnical challenges, based on the state-of-the-art practices, innovative technologies, new research results and case histories in construction and design towards safer and cost-effective infrastructures. This volume covers a wide range of topics with direct relevance to people within the broad field of civil and geotechnical engineers, including consultants, contractors, academics, materials suppliers and the owners and operators of civil infrastructures. Many papers associated with numerical modelling of transport infrastructure, advanced soil and rock testing, field monitoring, tunnelling, piles, geo-centre motion, triaxial experiments and dynamic testing are included. This volume specifically provides the following:

- Cutting-edge numerical and experimental solutions for sustainable civil infrastructures dealing with rocks and soils.
- New approaches and models to find soil behaviour, incorporating, nonlinearity of materials, dynamic and seismic response of granular materials.
- Principles, design, installation and monitoring aspects of deep foundations.
- Innovative technology to improve road safety

# Contents

# About the Editors

**Associate Professor Hadi Khabbaz** is Deputy Head of School for Research and Associate Professor in the School of Civil and Environmental Engineering, University of Technology Sydney (UTS). He received his Ph.D. in Geotechnical Engineering from the University of New South Wales, Australia. He is a recognised researcher in the area of pavement geotechnics, railway geotechnics, recycled waste applications and unsaturated soil mechanics. His research focuses on the theoretical and numerical analysis of soft soils, granular particles and unsaturated porous media. He has an H-Index of 33 in Google Scholar (28 in Scopus). He has 4100+ citations and 200+ publications. He was Deputy Chair and Chair of Australian Geomechanics Society (Sydney) in 2015-19. His awards include the Australian Learning & Teaching Council (ALTC) Citation Award, four best paper awards, UTS Learning & Teaching and Research Students Supervision Awards.

**Professor Jia-Ruey Chang** is Full Professor of the Graduate Institute of Architecture and Sustainable Planning at National Ilan University, Taiwan (ORCID: 0000-0001-8313-7325). He received his Ph.D. degree from the Department of Civil Engineering at National Central University, Taiwan. He currently serves as the President of Chinese Society of Pavement Engineering (CSPE), Editor-in-Chief of International Journal of Pavement Research and Technology (IJPRT), Springer, and Chairman of Pavement Engineering Committee of Chinese Institute of Civil and Hydraulic Engineering (CICHE). His research focus is on the application of advanced methodologies and technologies to pavement engineering and management. He has authored, co-authored and edited over 50 scientific journal papers, books and conference papers.

**Professor Yang Xiao** is a Full Professor of the School of Civil Engineering, Chongqing University. He received his Ph.D. degree from College of Civil and Transportation Engineering, Hohai University, Nanjing, China. He is a recognised researcher in the area of rockfill dam, railway geotechnics, bio-treatment geotechnics, recycled waste and granular flow mechanics. His research focuses on the testing and theoretical modelling of clay, silt, sand, rockfill, ballast and bio-cemented soils. He has an H-Index of 27 in Scopus. He has 2300+ citations and 100+ publications. He got many awards, such as 2018 and 2019 Most Cited Chinese Researchers from Elsevier, John Carter Award in IACMAG, Excellent Doctoral Dissertation of Chinese Society for Rock and Mechanics and Engineering, etc. He is also Associate Editor of International Journal of Geomechanics, ASCE, and Editor Board Member of Acta Geotechnica; Canadian Geotechnical Journal; Soils and Foundations, etc.

# Effect of Microscopic Properties on the Mechanical Behavior of Gravelly Soil by Using DEM

Ho Thi Kim Thoa and Meng Chia Weng(✉)

Department of Civil Engineering, National Chiao Tung University, Hsinchu, Taiwan
mcweng@nctu.edu.tw

**Abstract.** The gravelly soil has a wide range of grain size distribution, and its mechanical properties are much different from the sand, clay, or gravel solely. The amount and properties of fine content play an important role in the mechanical behavior of gravelly soil. To investigate the effect of fine content, this study collected a series of drained triaxial test data of gravelly soil under different confining pressure, ranging from 1.2 MPa to 3.0 MPa. Then, the discrete element method (DEM) was adopted to simulate the behavior of gravelly soil. The DEM simulation was first validated with the test data of gravelly soil. The effect of microscopic factors, including grain size, porosity, and particle properties, are further investigated in this study. PFC2D was adopted with the trial input parameters of the simulation. The existence of many fine particles affected the stress-strain relationship on simulation. The effective modulus is proportional to the magnitude of shear stress, while the contrary response with shear strain.

## 1 Introduction

The gravelly deposits usually distribute in the foothills, coastal, or river terrace areas in Taiwan. The thicknesses of gravelly formations range from 20 m to greater than 200 m. Gravelly deposits have complicated mechanical properties because they contain large gravel particles and cemented matrix (Fig. 1), in which the particles coarser than 4.75 mm are regarded as gravel, and the matrix represents the particles finer than 0.075 mm. Remarkably, the grain size distribution of this soil type is quite wide with the diameter of gravel is up to hundreds of millimeters. These characteristics have significantly affected the mechanical properties of gravelly soils. Thus, the mechanic properties of these soils are very different from other materials such as sand, clay, or gravel individually, and they may influence the excavation, tunneling, and foundations such as squeezing and collapse (Chu et al. 1995; Chu et al. 1996; Weng et al. 2013).

Most engineering practices assumed the gravelly deposits as none-cohesion and range of friction angles between 32° to 35°, which often underestimate the shear strength of gravelly soil (Chang et al. 2015). Therefore, the investigation of mechanical properties of gravelly soils is necessary for foundation design. The large-scale triaxial test is suitable to evaluate the behavior of gravelly soil. However, the preparation of large specimens and the experimental procedure are complicated and difficult so that the number of


H. Khabbaz et al. (Eds.): GeoChina 2021, SUCI, pp. 1–14, 2021.
https://doi.org/10.1007/978-3-030-79650-1_1

(a) The gravelly soil in Taichung, Taiwan (b)The gravelly soil in Kaohsiung, Taiwan

**Fig. 1.** The gravelly soil in some areas of Taiwan

studies related to the mechanical properties of gravelly soil is much less than other soils. To realize the effect of microscopic properties on the mechanical behavior of gravelly soil, this study adopted a discrete element method to simulate the behavior of gravelly soil. Base on the results of the large-scale triaxial test with pure shear stress path, this study first compared the simulation and the experimental results. Then, the microscopic characteristics of gravelly soil, such as grain size, initial porosity, particle stiffness, and interparticle friction coefficient was analyzed. The major influencing factor was further clarified.

## 2 Triaxial Tests on Gravelly Soils

### 2.1 Specimen Preparation

In this study, a series of triaxial test data of gravelly soils were collected. The gravelly soil samples were taken from central Taiwan. A series of remolded specimens were prepared to obtain the mechanical properties of gravelly soils. The grain size distribution curves of these samples are based on the natural deposit. However, the maximum grain size was scaled down to 5.1 cm because of the testing equipment limitation (Fig. 2) (Chu et al. 2010; Weng et al. 2013). These types of remolded soils are classified as poorly graded gravel with silt (GP-GM) through the Unified Soil Classification System (USCS; ASTM 2006) and A-1-a through the AASHTO classification system (Chu et al. 2010; Weng et al. 2013). The fine content (silts and clays) of these remolded samples is approximately 8.7%.

Firstly, these remolded specimens were compacted as standard Proctor compaction energy, the maximum density and the optimum moisture content (OMC) got 2120 kg/m$^3$ and 5.9%, respectively. After that, the large specimens with 30.5 cm in diameter and two times of diameter (610 mm) in height were prepared for the triaxial test. Before conducting the test, these specimens were saturated at least 36 h to reach the pore-pressure parameter, B = 0.95 under a back pressure of 0.3 MPa.

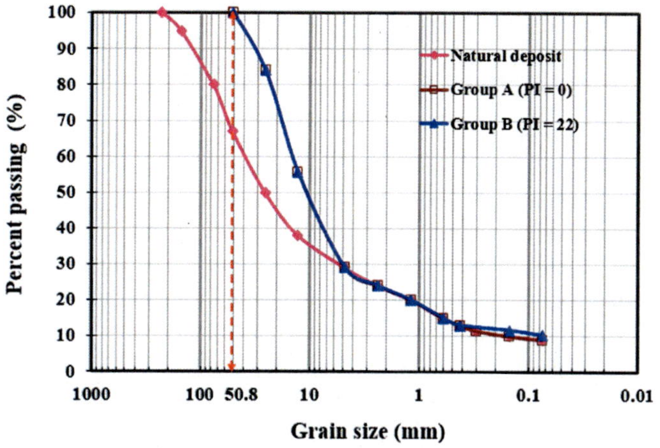


**Fig. 2.** Grain size distribution curves of natural deposit and remolded specimens (Chu et al. 2010)

## 2.2 Experimental Setup

A series of drained triaxial compression (CID) tests were carried out to investigate the mechanical behavior of gravelly soils. The axial load was provided by a loading frame which achieved the maximum capacity of 500 kN with the loading rates being 0.001–30 mm/min. The hydrostatic pressure was monitored by a pressure transducer with an accuracy of 0.001 MPa. The maximum pressure capacity of the triaxial cell was 5 MPa. Three stress paths (Fig. 3) were conducted by using servo control both axial and hydrostatic stresses. In these triaxial tests, the axial deformation and the volume change of specimens were measured by the electric displacement gauge and the water volume change meter with an accuracy of 0.1 mm and 0.1 ml, respectively.

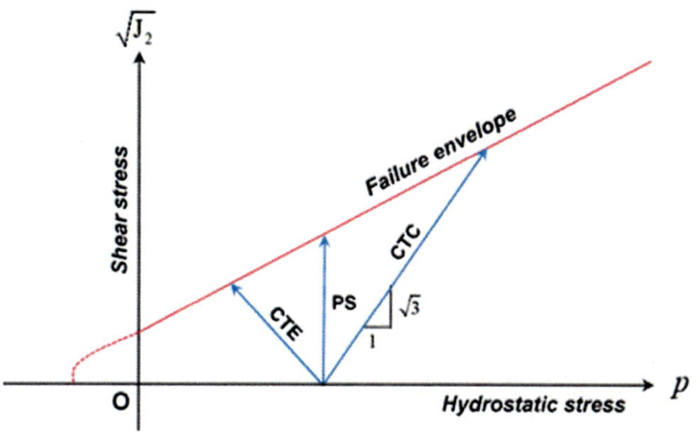


**Fig. 3.** Schematic illustration of the conducted stress paths. PS, pure shear test; CTC, conventional triaxial compression test; CTE, conventional triaxial extension test (Chu et al. 2010; Weng et al. 2013)

# 3 DEM Model Simulation

## 3.1 Model Generation

This study used the Particle flow code in two dimensions (PFC2D) to simulate the abovementioned tests. The model of the triaxial test with pure shear stress path was simplified as a two-dimensional sample, within the trial method input parameters. The results of this study are the foundation for the next three-dimensional research.

The particles of gravelly soil were simulated by the ball particles with various diameters. Considering the computing efficiency, it is impossible to simulate the triaxial test by using the actual particle sizes, so this study chose the particle size ranging from 0.3 mm to 50.8 mm according to the grain size distribution curve of experimental samples (Fig. 4). In the calibration distribution grain size, the percentage of size particles less than 0.3 mm will be counted into grain size 0.3 mm, which was regarded as fine content in DEM analysis.

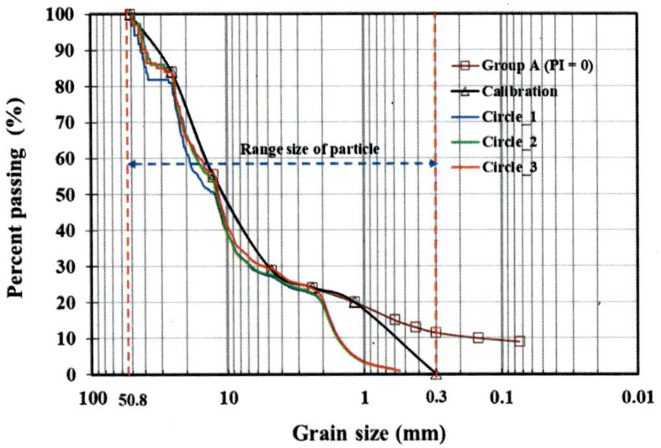


**Fig. 4.** Grain size curves of experimental soil and measured samples

The specimen of numerical analysis was 610 mm high and 305 mm wide, the same as the size of the actual specimen. The specimen was confined by four separate walls, and the upper and bottom walls were regarded as the top and bottom plates in the load frame. To reduce the effect of boundary during the preparation, the wall friction was set to zero, so that the shear force could not develop in facet-ball contacts. The particles were generated in the specimen range with a target porosity of 12.5% which was measured from the laboratory results. A total of 24,422 particles were filled in the specimen to achieve the target porosity, (Fig. 5). When the particles were randomly distributed, large gaps appeared in the specimen (Fig. 5a). Thus, the interparticle sliding could take place in a relatively effortless manner when a small frictional force was applied to grain contacts during the specimen generation stage to generate a dense specimen (Yang et al. 2018) (Fig. 5b). At the equilibrium state, the porosities of n = 11.8%, n = 12.1%, and n = 12.4% were measured by three circles in Fig. 5c, respectively.

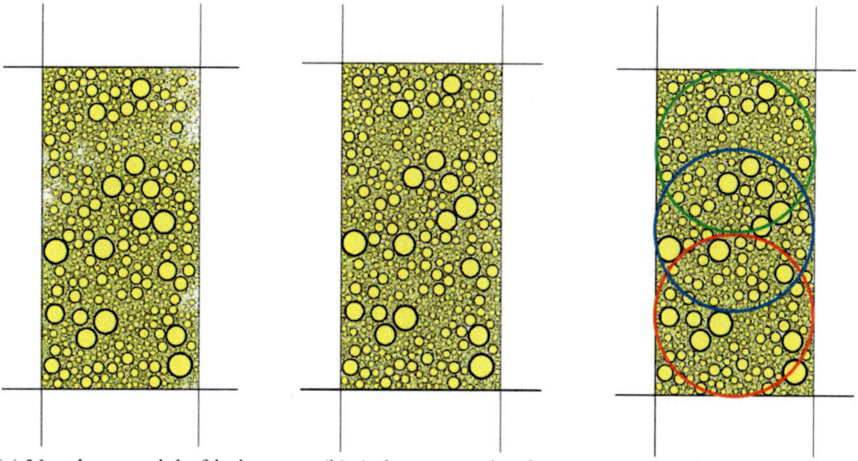

(a) Non-interparticle friction equilibrium sample (b) A dense sample after equilibrium state (c) The measured circles for size distribution calculation

**Fig. 5.** The states of the sample after solving to get the target average ratio

## 3.2 Determination of Micro Parameters

To consider the contact behavior between particles, the linear contact model was adopted based on the linear deformability. This model was described by effective modulus ($E^*$) which related to the elastic constants of Young modulus ($E$) and the normal to shear stiffness ratio ($k^* = k_n/k_s$) which related to the Poisson's ratio ($\nu$) based on the following equations for a 2D model (Itasca 2016). This model has been used in many previous studies in simulating the triaxial tests to determine the properties of granular materials under various loading conditions.

$$k_n = \frac{AE^*}{L} = \frac{2[\min(R^{(1)}, R^{(2)})]E^*}{(R^{(1)} + R^{(2)})}; \text{ for the ball} - \text{ball} \tag{1}$$

$$k_n = \frac{AE^*}{L} = 2E^*; \qquad \text{for the ball} - \text{wall} \tag{2}$$

$$k_s = \frac{k_n}{k^*} \tag{3}$$

The required microscopic parameters for linear contact models are summarized in Table 1.

**Table 1.** Summary of input parameter for gravelly soil triaxial test

| Parameters | Unit | Value |
|---|---|---|
| Contact model | | Linear contact model |
| Height × diameter | mm | 610 × 305 |
| Number of particles | | 24422 |
| Particle friction coefficient, μ | | 0.72 |
| Wall friction coefficient | | 0 |
| Effective modulus, $E^*$ | Pa | 2.8 × 108 |
| Normal-shear stiffness ratio, $k^*$ | | 2.5 |
| Density of particles, $\gamma_d$ | kg/m$^3$ | 2150 |
| Porosity of specimen | | 0.125 |
| Max. diameter of particles | mm | 50.8 |
| Min. diameter of particles | mm | 0.3 |

### 3.3 Modeling Procedure

This study simulated a series of triaxial tests of gravelly soil under different confining pressure ($\sigma_3$), 1.2 to 3.0 MPa. The procedure of simulation processing is illustrated in Fig. 6.

The procedure of the triaxial test could be divided into two stages: (1) Isotropic consolidation stage and (2) Pure shear stress path stage.

In the isotropic consolidation stage, the specimen was compressed isotropically under a prescribed confining pressure through the wall servo mechanism. To apply target stress for all walls, the force was applied from the wall servo command. The specimen was considered to be equilibrated when the tolerance of stress obtained from the walls was less than 0.5%, and the ratio of the mean static unbalanced force to the mean contact force was less than $10^{-5}$ (Hazzar et al. 2020). In the pure shear stress path stage, the axial stress continued to increase while the lateral stress decreased. The hydrostatic pressure was kept as a constant by decreasing the cell pressure to 50% of the increment of the axial stress ($\Delta\sigma_2 = \Delta\sigma_3 = -0.5\Delta\sigma_1$). In this model, the pure shear stress path was imitated by using a wall servo mechanism until the lateral stress equal 50% of confining pressure.

During the processing, the stresses were obtained from the contact forces impacted on the walls and the boundaries of the specimen. The strains were calculated from the displacement of the walls directly.

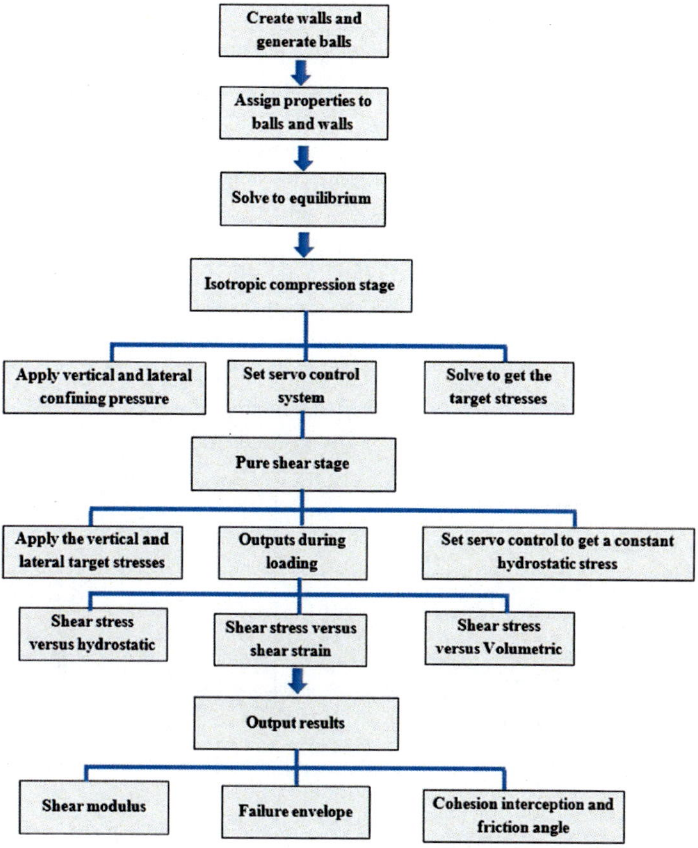


**Fig. 6.** The procedure of triaxial test simulation by DEM.

The shear stress ($\sqrt{J_2}$), shear strain ($2\sqrt{J_2'}$), and volumetric strain ($\varepsilon_v$) were determined by the following equations:

$$\sqrt{J_2} = \frac{1}{\sqrt{3}}(\sigma_1 - \sigma_3) \tag{4}$$

$$\gamma = 2\sqrt{J_2'} = \frac{2}{\sqrt{3}}(\varepsilon_1 - \varepsilon_3) \tag{5}$$

$$\varepsilon_v = \varepsilon_{kk} = \varepsilon_1 + 2\varepsilon_3 \tag{6}$$

## 4 Simulation Results

### 4.1 Deformation Induced by Shear Stress

Figure 7 shows the comparison of shear strain and volumetric strain induced by shear stress through the pure shear stress path (path AC in Fig. 3) from the simulation and experimental results. The simulations agree well with the experimental stress-strain curves. Under the shear stresses, the shear strains are relatively linear initially. The significant increase in shear strain occurred when the stress path reached the failure state (Fig. 7a). The volumetric strains caused by shearing are compressive at the early stage, and the volumetric strain converts to dilation as the shear stress increases (Fig. 7b). The simulation under different hydrostatic pressure provides good agreement with the test results, increasing the confining pressure can enhance the material strength. However, the result still has a little discrepancy in the magnitude of shear strains. Theoretically, the hydrostatic stress in the pure shear stress path is constant during the pure shear stage. The results, presented in Fig. 7c show that the stress paths were slightly curvy, but they were still reasonable. It can be seen that these errors are caused by excessive compression of the sample after the consolidation stage.

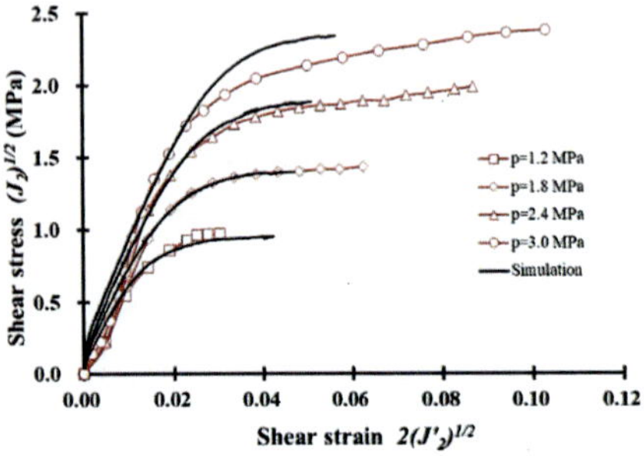


(a) Shear stress versus shear strain

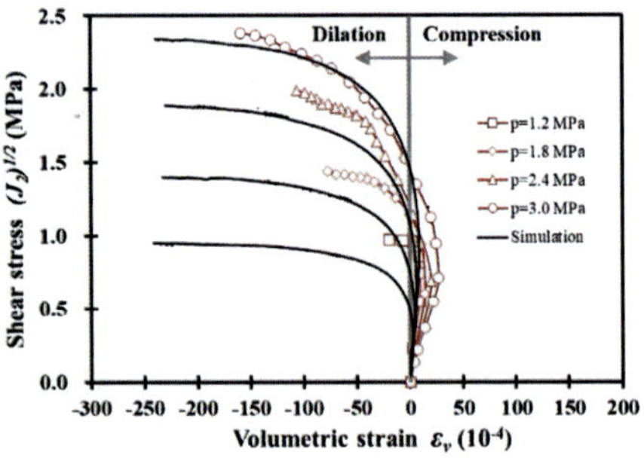


(b) Shear stress versus volumetric strain

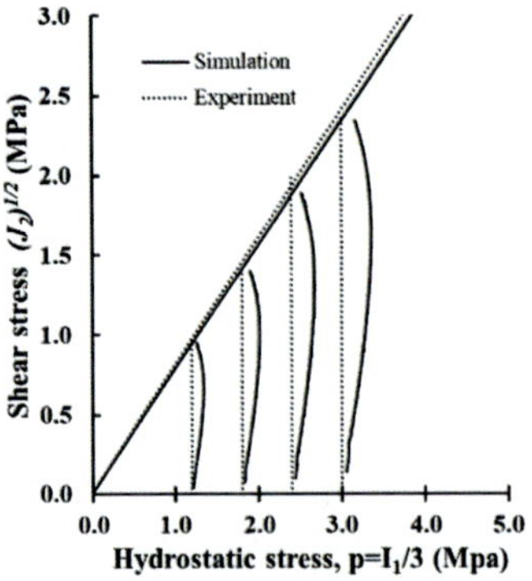


(c) The pure shear stress paths under various hydrostatic stresses

**Fig. 7.** The comparison of simulated and actual deformations of gravelly soils.

## 4.2 Failure Envelope

The failure envelopes in the case of experimental and simulated results are shown in Fig. 8. These envelopes were obtained from the maximum shear strength of tests. Based on the Drucker-Prager criterion, the failure envelope could be expressed by the relationship of shear strength ($\sqrt{J_{2,f}}$) and first stress invariants of the three-dimensional stress components.

$$\sqrt{J_{2,f}} = \alpha I_1 + k \tag{7}$$

where α and $k$ are the slope and cohesive intercept of the failure envelope, respectively. As a result, the slope α and the intercept $k$ of the failure envelope in this comparison are consistent. The failure envelope expressed by Eq. (7) can be converted to the Mohr-Coulomb failure criterion. The values of friction angle ϕ and cohesive intercept $c$ are compared in Table 2.

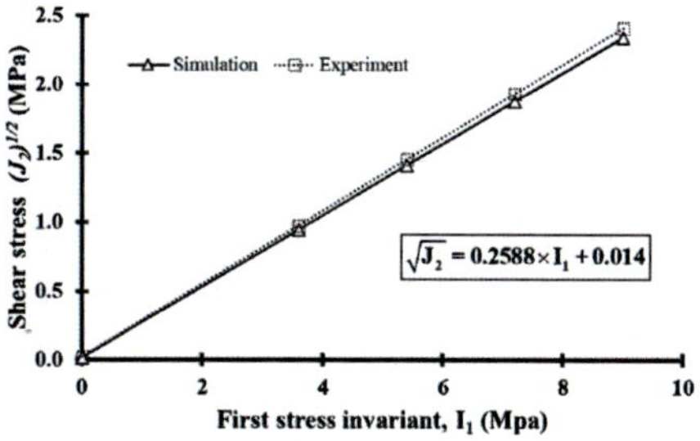


**Fig. 8.** The failure envelope of the tested and simulated gravelly soil

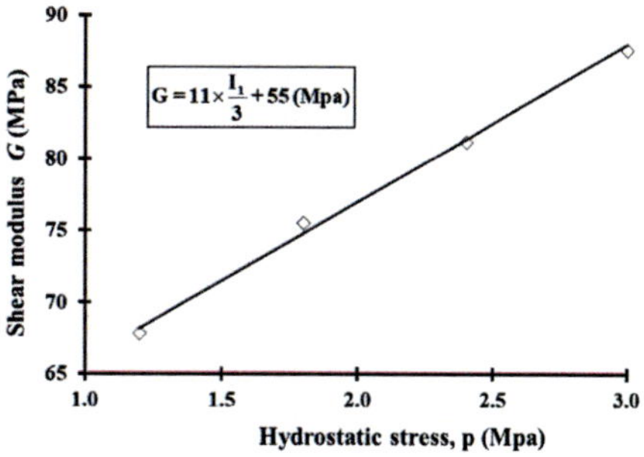


**Fig. 9.** The simulated shear modulus under various hydrostatic stresses

The elastic shear modulus (G) was determined based on the ratio of shear stress ($\sqrt{J_2}$) and elastic shear strain in ($2\sqrt{J_2'}$). The shear stress was calculated at 50% of the failure shear stress.

$$G = \frac{\sqrt{J_2}}{2\sqrt{J_2'}} \tag{8}$$

The elastic shear modulus in this simulation tends to increase linearly when expressed for different confining pressures. As a result, the shear modulus ranges from 68 to 88 MPa (Fig. 9).

**Table 2.** The macro parameters of soils obtained by simulation

| Parameters | Experiment | Simulation |
|---|---|---|
| Friction angle (degree) | 34.2 | 33.32 |
| Cohesive interception (MPa) | 0.016 | 0.014 |

# 5 Effect of Micro Parameters on the Performance of the Simulation

## 5.1 Effect of Porosity

The different initial porosity, ranging from 0.05 to 0.3, was considered in the simulating. Under the same confining pressure rate (p = 1.8 MPa), a larger initial porosity will lead to a larger volumetric strain at the end of the consolidation stage (Fig. 10a). For gravelly soil, the size distribution is wide, and the increasing hydrostatic pressure caused the small particles to move easily, which generates a larger volumetric strain. Due to the high fine content in the specimen, the volumetric strains deform significantly during the consolidation stage. The relationship between the increments of volumetric strain and initial porosity is nonlinear. It could be described as a power function equation of initial porosity. After the hydrostatic compression stage, the compacted porosity of the specimens was much less than the initial porosity, which means the main compression was generated during the consolidation stage. In general, the compacted porosity of specimens was approximately 20% less than the initial porosity. The less initial porosity, the less compacted porosity (Fig. 10b).

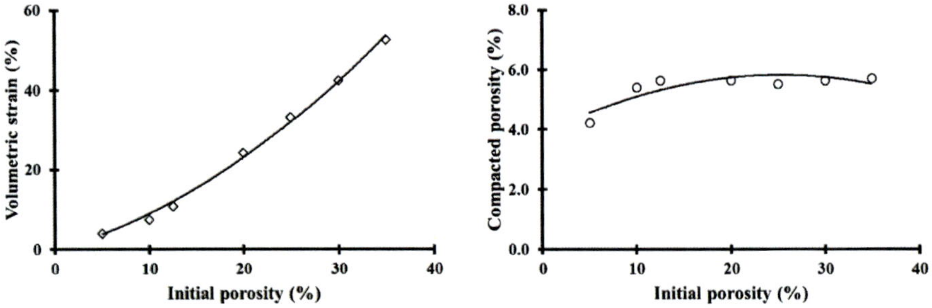


(a) Volumetric strain curve under initial porosity

(b) The compaction of soil sample

**Fig. 10.** The effect of void rate in the consolidation stage

A smaller initial porosity induced the higher shear strength and the less shear strain and dilative volumetric strain (Fig. 11).

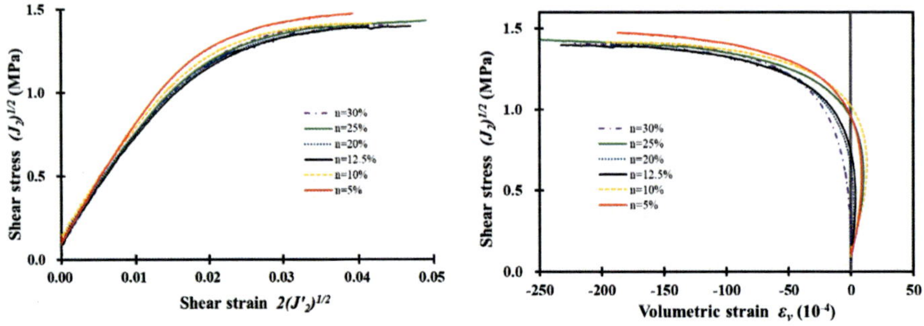


**Fig. 11.** The effect of void rate in the pure shear stage

## 5.2 Effect of Particle Size

In this part, the upper grain size was kept fixed (50.8 mm), the minimum grain size was changed to investigate the effect of particle size on the mechanical behavior. The correlation between the minimum size of the particle is nonlinear negatively correlated with the volumetric strain (Fig. 12a), and proportional linearly correlated with compacted porosity (Fig. 12b). For the sample had a wide range size of particles, after the first stage the volumetric strain was larger but the compacted porosity was less.

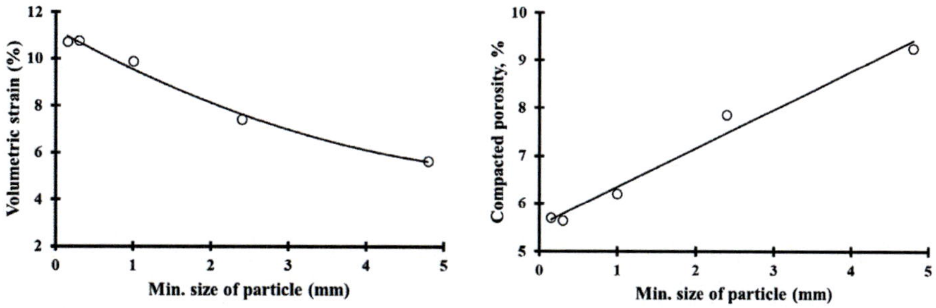


(a) Volumetric strain curve under initial porosity

(b) The compaction of soil sample

**Fig. 12.** The effect of the minimum size of particles in the consolidation stage

The influence of the particle size is insignificant on the shear strength and strain in this simulation. For cases of small lower grain size, the differences of shear stress, shear strain, and volumetric strain are not significant (except 4.8 mm) (Fig. 13).

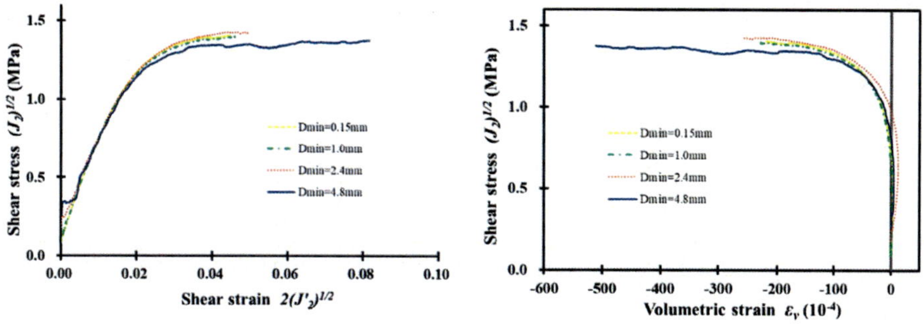


**Fig. 13.** The effect of the minimum size of particles in the pure shear stage

## 5.3 Effect of Effective Modulus

In the contact model, effective modulus ($E^*$) is one of the critical input parameters of this model. The effective modulus affects not only the stress but also the strain in the pure shear. With shear stress–shear strain curve, the slope of the curve increases gradually with the increase of effective modulus. It can be noted that the shear modulus (Eq. 8) is in proportion with effective modulus. On the contrary, the smaller effective modulus generates a larger volumetric strain on dilation (Fig. 14).

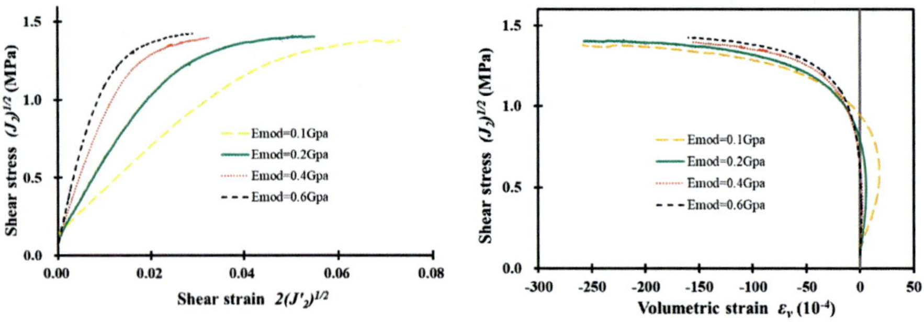


**Fig. 14.** The effect of the effective modulus

## 5.4 Effect of Normal – Shear Stiffness Ratio

Simulations with a variety of normal to shear stiffness ($k^*$) was conducted to point out the impact of this coefficient on the triaxial test on gravelly soil. Figure 16 shows the variations of shear strength, shear strain, and volumetric strain under different $k_n/k_s$ ratio. As a result, the value of $k^* = 1.0$ has the highest shear strength, the smallest shear strain, and volumetric strain, while the case of $k^* = 10$ exhibits an opposite tendency (Fig. 15).

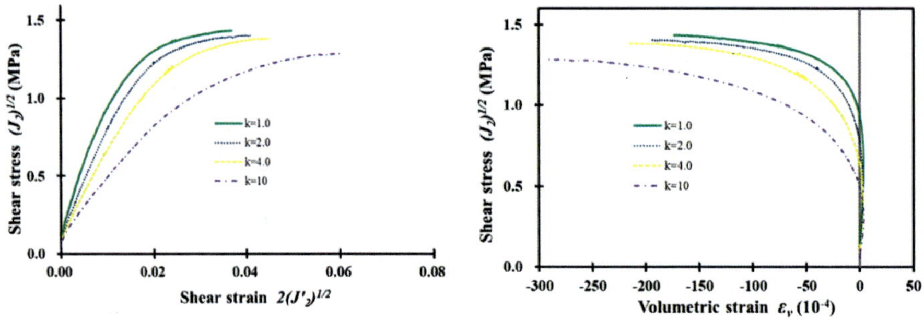


**Fig. 15.** The effect of the normal to shear stiffness ratio

In addition to this, the normal to shear stiffness also affect the stress path in simulation. The smaller $k^*$ value, the higher stability of the simulated stress paths (Fig. 16).

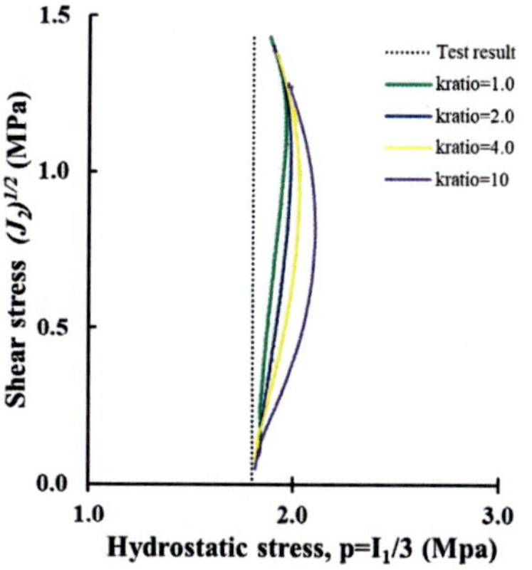


**Fig. 16.** The effect of $k_n/k_s$ to stress path.

## 5.5 Effect of Interparticle Friction Coefficient

The friction coefficient of the particle appears to only affect the peak response (Potyondy et al. 2004). This study provided similar results, which agree with the previous studies in the investigation of the interparticle friction coefficient (Fig. 17).

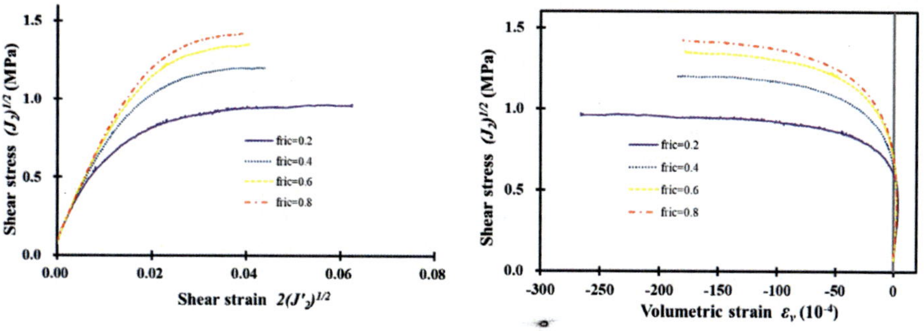


**Fig. 17.** The peak of stress under various interparticle friction coefficient

# 6 Conclusion

The objective of the study is to evaluate the effect of fine content in gravelly soils on their mechanical properties using DEM software PFC2D. The finer content which was represented by the grain size distribution with more than 24% of fine particles. The numerical triaxial tests for the pure shear stress path corresponds to the confining pressure 1.2, 1.8, 2.4, and 3.0 MPa respectively. A series of simulation of changing the input parameters were conducted.

1. The simulation could reveal the mechanical behavior of gravelly soil, including the shear strength, shear strain, and volumetric strains. The simulated failure envelope of

the studied gravelly soils almost coincides with the experimental results. The friction angle $\phi$ and cohesion intercept $c$ are 34.2° and 0.016 for test results and 33.32° and 0.014 for simulations.

2. The initial porosity and the fine content of the modeled sample primarily affect volumetric strain during the consolidation stage. In this stage, the volume of the sample was greatly deformed under large initial porosity, also much fine content. However, the effect of initial porosity on stress-strain curves is ambiguous with a value of the initial porosity is more than 20%.
3. The study considered the effect of micro properties on the behavior of gravelly soils. The results showed that the interparticle friction coefficient only related to the peak of the strength of soil; the effective modulus had a relative impact on deformation in an inverse proportion to shear strain and a proportional rate to volumetric strain; the normal to shear stiffness ratio controlled the failure shear stress in negative correlation, shear strain and dilative volumetric strain in positive correlation.
4. Remarkably, all simulations in this study were investigated in 2D, while the test results in 3D. Some functions and calculations were calibrated to achieve the approximate results. Besides, the effect of particle shape has not been considered in the simulation.

## References

de Bono, J.P., McDowell, G.R.: DEM of triaxial tests on crushable sand. Granular Matter **16**(4), 551–562 (2014). https://doi.org/10.1007/s10035-014-0500-x

Chang, K.T., Kang, Y.M., Ge, L., Cheng, M.C.: Mechanical properties of gravelly deposits evaluated by nonconventional methods. J. Mater. Civ. Eng. **27**(11), 04015032-1-04015032–11 (2015)

Chu, B.L., Jou, Y.W., Weng, M.C.: A Constitutive model for gravelly soils considering shear-induced volumetric deformation. Can. Geotech. J. **47**(6), 662–673 (2010)

Potyondy, D.O., Cundall, P.A.: A bonded-particle model for rock. J. Rock Mech. Min. Sci. **41**, 1329–1364 (2004)

Hardin, B.O., Kalinski, M.: Estimating the shear modulus of gravelly soils. J. Geotech. Geoenviron. Eng. **131**(7), 867–875 (2005)

Hazzar, L., Nuth, M., Cherkired, M.: DEM simulation of drained triaxial tests for glass-beads. J. Powder Technol. **364**, 123–134 (2020)

Weng, M.C., Chu, B.L., Ho, Y.L.: Elastoplastic deformation characteristics of gravelly soils. J. Geotech. Geoenviron. Eng. **139**, 947–955 (2013)

Yang, D., Behzad, F., Hadi, K., Henry, Z.: Influence of particle contact models on soil response of poorly graded sand during cavity expansion in discrete element simulation. J. Rock Mech. Geotech. Eng. **10**, 1154–1170 (2018)

# Research on Large Deformation Control Measures of Surrounding Rock in Jointed and Altered Granite Stratum of High Ground Stress Tunnel

Xinghua Fang(✉), Junsheng Yang, Xuemin Zhang, Linyi Li, and Weilong Liu

School of Civil Engineering, Central South University, Changsha, Hunan, China
fangxinghua@csu.edu.cn

**Abstract.** The large deformation of surrounding rock occurred frequently when Zangga tunnel passed through the high-stress jointed altered granite formation, and the deformation could not be controlled by the design scheme proposed in the initial stage of construction. Based on the monitoring results and geological data, this paper analyzed the deformation characteristics of jointed and altered granite in Zangga tunnel, and discussed the influencing factors of deformation. By means of a field test, the comprehensive control measures of surrounding rock deformation, which mainly includes adjusting the curvature of sidewall, combining long and short bolts and double-layer primary support, were studied. The results show that the deformation is characterized by large deformation value, fast deformation rate and long duration, and its tunnel convergence is greater than crown settlement. The influences of active fault on surrounding rock and alteration of surrounding rock are the internal causes of large deformation of the tunnel, while high ground stress is the external factor. The field test results show that the deformation value, rate and duration of surrounding rock can be controlled after the comprehensive control measures are adopted, and the construction efficiency is greatly improved.

**Keywords:** Altered granite stratum · High ground stress · Deformation characteristics · Double-layer primary support · Control measures

## 1 Introduction

The altered rock formed by the influence of magma intrusion is quite different from the original rock in mineral composition and structure. The altered rock usually has low density, strength and deformation modulus, and the rock mass is relatively broken. When the tunnel crosses the altered rock construction, problems such as large deformation, collapse and water gushing are prone to occur.

As a typical hard metamorphic rock, the density, strength and other physical and mechanical properties of altered granite are quite different from its original rock, and it is characterized by weak rock mass and poor self-stability. Scholars have carried out the following aspects of research: Fracture development and failure process of altered granite in uniaxial compression test (Coggan et al. 2007); Differences in physical and mechanical


H. Khabbaz et al. (Eds.): GeoChina 2021, SUCI, pp. 15–28, 2021.
https://doi.org/10.1007/978-3-030-79650-1_2

properties of granite under weathering and hydrothermal alteration (Khanlari and Naseri 2016); Hydrothermal alteration zone identification method and its relation to seismic activity (Meller and Kohl 2014). Previous studies mainly focus on the physical and mechanical properties of altered granite, but the study on the deformation characteristics of surrounding rock and supporting methods of the tunnel through altered granite is still not mature. Therefore, it is necessary to carry out relevant research.

Based on the monitoring results of jointed and altered granite stratum of Zangga tunnel, this paper analyzes the deformation characteristics of surrounding rock and discusses the influencing factors of deformation combining with geological and construction data. And the control measures of large deformation of surrounding rock is studied by field test.

# 2 Project Overview

## 2.1 Engineering Background

Zangga Tunnel is located between Nyenchenthanglha and Himalayas. The tunnel area is characterized by high mountains and deep valleys, rugged terrain, extreme harsh climate and typical mountain and canyon landforms. As one of the control projects of the Lhasa-Linzhi Railway, Zangga tunnel is a passenger and cargo collinear single-track tunnel with a total length of 8.755 km and a designed train speed of 160 km/h. The entrance mileage of Zangga tunnel is DK164+850, the exit mileage is DK173+605, the ground elevation is 3550–4400 m, and the maximum burial depth is about 778 m. The location and plane layout of Zangga tunnel is shown in Fig. 1.

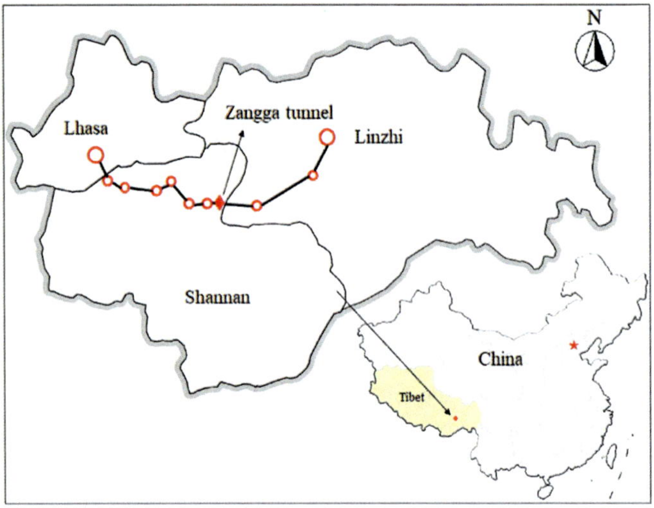


**Fig. 1.** The location and plane layout of Zangga tunnel

## 2.2 Engineering Geology

The neotectonic movement in the tunnel area is strong, and the regional fault structure is extremely developed. The rock mass is relatively broken under the influence of regional structure, and is affected by magma intrusion for many times. The local rock mass alteration characteristics are obvious. The DK168+805–DK169+140 section of the tunnel mainly passes through weakly weathered granite ($E_2R$). The rock mass has developed joints and fractures with a steep dip Angle. The local rock mass has a low strength under the influence of alteration and poor self-stability ability. The DK168+805–DK169+140 section of the tunnel mainly passes through weakly weathered granite ($E_2R$). The rock mass has developed joints and fractures with a steep dip Angle. The local rock mass has a low strength under the influence of alteration and poor self-stability ability. The groundwater in the rock mass in this section is moderately – weakly developed, and there is angular unconformity contact among the rock masses. The longitudinal section diagram of typical tunnel sections is shown in Fig. 2.

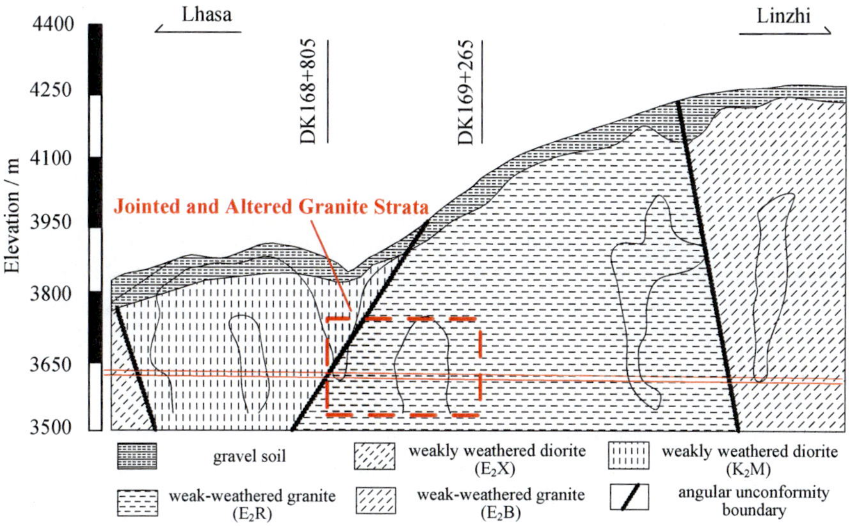


**Fig. 2.** A longitudinal section of a typical section in Zangga tunnel area

By measuring the ground stress of the deep hole in the tunnel area, it can be seen that the average ratio of the maximum horizontal principal stress and vertical stress is 1.29 and the maximum value is 1.76. This indicates that the ground stress in the tunnel area is mainly tectonic stress, and its direction is N6°W–N7°E. And the maximum horizontal ground stress is 17.72 MPa.

## 2.3 Situation of Tunnel Construction

On March 13, 2017, construction of the tunnel began at DK169+000, where the tunnel intersects with the inclined shaft. The surrounding rocks revealed by excavation are jointed and altered granite, as shown in Fig. 3. The rock mass has developed joints and fractures with a steep dip Angle, and the structural plane is infected with iron and manganese, with visible scratches and the development characteristics of fault gouges.

The rock mass has low strength and poor integrity, the partial rock mass is easily broken by hand, and surrounding rock has a poor self-stability ability.

**Fig. 3.** Jointed and altered granite

The surrounding rock grade and the original support design scheme were changed on-site because the surrounding rock was exposed to be broken. DK169+025–DK169+140 section of the surrounding rock grade by III level changed for V level, and the construction method by entire section method changed to bench method. After design change, the segment most areas supporting structure adopted $V_c$ composite lining, which section height was 10.15 m and width was 8.30 m, as shown in Fig. 4. In the primary support, anchor rods with a length of 3 m and a diameter of 22 mm are set at the arch and sidewall positions, with a circular spacing of 1.2 m and a longitudinal spacing of 1.0 m. In addition, the spacing of section steel frames is 0.8 m, and the thickness of C25 shotcrete is 250 mm. The reserved deformation value of tunnel primary support is 100 mm. And the secondary lining is made of 450 mm thick C35 reinforced concrete.

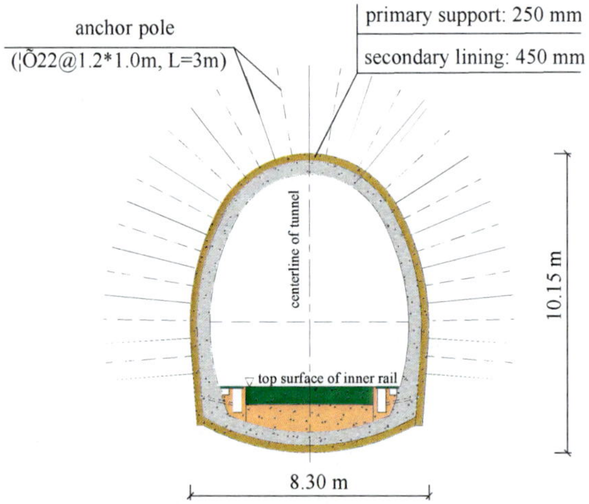


**Fig. 4.** Section view of $V_c$ composite lining structure

The construction sequence of DK169+025–DK169+140 are as follows: ① The advanced reinforcement of the arch of the tunnel face. ② Excavate and support the upper bench with a height of 4.4 m and a length of 10 to 20 m. ③ The left and right sides of the bottom bench are staggered apart from the excavation and support, the bottom bench is 4.2 m high, and each bench is 3–8 m long. ④ Excavate and support the tunnel invert. ⑤ Complete the secondary lining according to the monitoring measurement results. Construction sequence of bench method is shown in Fig. 5.

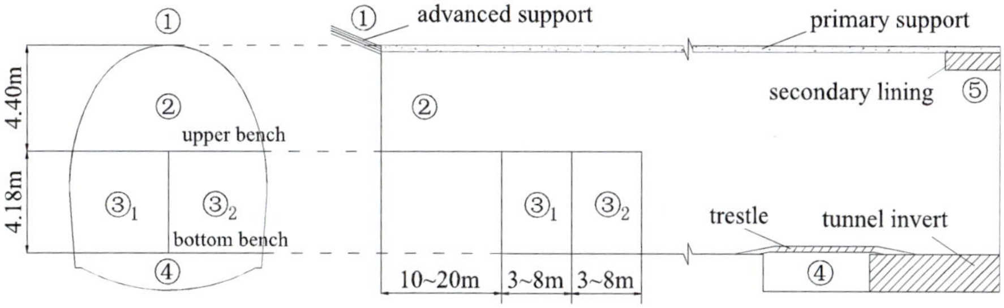


**Fig. 5.** Construction sequence of bench method

With the progress of construction, large deformation of surrounding rock appeared in different degrees in the tunnel area. Problems such as bending fracture of steel arch, cracking and dropping of jetted concrete, and deformation limit of primary support (shown in Fig. 6) occurred frequently. The maximum crack width of jetted concrete was 26 mm. For these problems, the construction unit removed and replaced the primary support, and these measures seriously affected the construction period. The deformation control measures of surrounding rocks suitable for this section needed to be further studied.

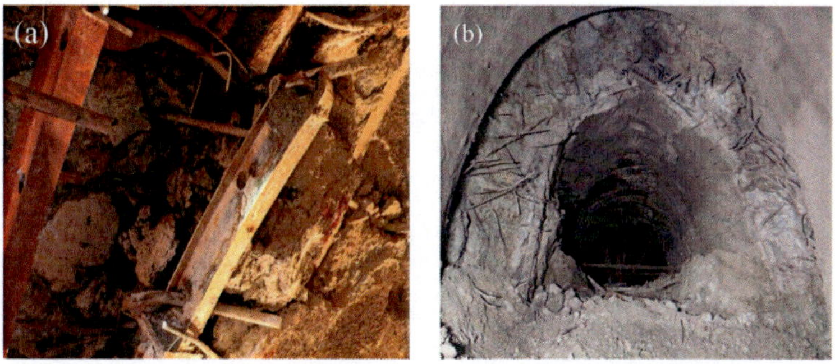


**Fig. 6.** Typical pictures of primary support damage: (a) fracture of steel arch; (b) schematic diagram of primary support deformation

# 3 Deformation Characteristics of Surrounding Rock and Its Influencing Factors

## 3.1 Overall Deformation of Large Deformation Section

Section DK169+025–DK169+140 is taken as the large deformation section for analysis. According to the data obtained from statistical monitoring, the surrounding rock deformation of this section is shown in Fig. 7. The measured point of tunnel convergence is located near the junction of upper and bottom bench, and the negative value represents the inward deformation of the tunnel. The crown settlement of this section is from 71 to 656 mm. The maximum daily settlement, which is located at DK169+122, is 35 mm; The tunnel convergence is from 245.3 to 3032 mm, and the maximum daily tunnel convergence, located at the junction of upper and bottom bench of DK169+090, is 128 mm. The maximum crown settlement and tunnel convergent deformation appear at section DK169+122. The tunnel convergent deformation of DK169+025–DK169+140 section is obviously greater than the crown settlement. Most of the section deformation exceeded the reserved deformation value (100 mm), and the primary support deformation seriously exceeds the limit value. Results show that the $V_c$ composite lining structure on the formation adaptability is poorer, and it cannot control the deformation of the surrounding rock.

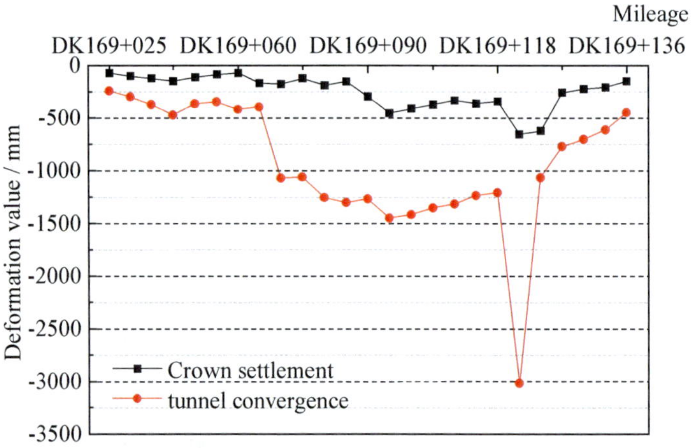


**Fig. 7.** Cumulative deformation of surrounding rock at the section of DK169+025–DK169+140

## 3.2 Deformation Characteristics of Surrounding Rock with Typical Section

Bench method was used for the DK169+090 section construction, and supporting structure adopted $V_c$ composite lining. The tunnel deformation and rate curves are shown in Fig. 8.

Within 48 days, the cumulative crown settlement value of this section is 295 mm, and the cumulative tunnel convergence value is 1272 mm. The deformation of surrounding rock is divided into the acceleration stage and development stage during the whole monitoring period, and the deformation keeps increasing.

After upper bench excavation, surrounding rock deformation increased rapidly, arch crown settlement was 13 mm within 1d excavation, and tunnel convergence was 90 mm. Subsequently, the crown settlement deformation kept a steady increase at a rate of 10–13 mm/d. After a short decrease, the tunnel convergence rate increased to a peak of 128 mm/d at 5d of excavation. After five days, the deformation rate of surrounding rock gradually decreased. Before excavation on the left side of the bottom bench, the crown settlement value reached 133 mm, accounting for 33.71% of the total deformation. The tunnel convergence value reached 704 mm, accounting for 55.33% of the total deformation. The deformation rate fluctuated in this stage.

After excavation on the left side of the bottom bench, the deformation rate of surrounding rock slowed down to a certain extent, but the deformation continues to increase at a steady rate. There was no obvious convergence trend in surrounding rock deformation after supporting tunnel invert.

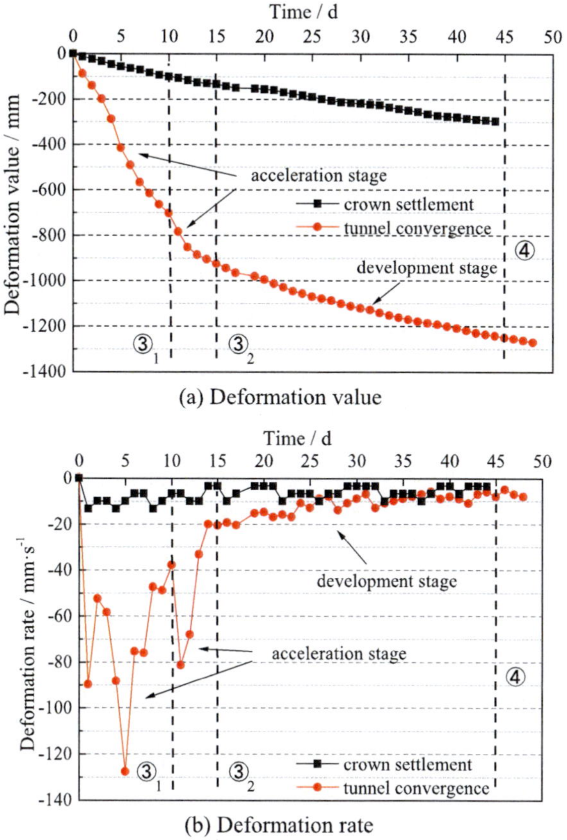


**Fig. 8.** The tunnel deformation and rate curves at the section DK169+090

### 3.3 Deformation Characteristics of Surrounding Rock

Combined with the deformation of typical sections, it can be seen that during the construction of section DK169+025–DK169+140, the surrounding rocks show the following deformation characteristics: ① The deformation of surrounding rock is characterized by large value and fast deformation rate, and the tunnel convergent deformation is obviously greater than the crown settlement deformation; ② The deformation of surrounding rock is divided into acceleration stage and development stage. After the support structure is closed, the deformation rate of surrounding rock does not decrease significantly and the deformation of surrounding rock continues to increase. ③ Construction has a great influence on the deformation of surrounding rock. During the excavation of upper and bottom bench, the deformation rate fluctuation of surrounding rock will occur, which has a great influence on the deformation value of surrounding rock. ④ The deformation of surrounding rock lasts for a long time. In 45 days, the deformation of surrounding rock shows no obvious convergence trend, and the deformation still keeps a slow growth.

### 3.4 Influencing Factors of Surrounding Rock Deformation

Based on the comprehensive analysis of surrounding rock deformation monitoring results and geological data, it is found that the main factors affecting surrounding rock deformation are as follows:

(1) Geological reasons
Fault zone distribution diagram of Zangga tunnel is shown in Fig. 9. The F5-2 fault in the eastern margin of Oiga Basin is a seismogenic fault of the Sangri 7.0-magnitude earthquake in 1915, with frequent strong seismic activity ($Q_4$ active fault). The tunnel area is located in the fault-affected zone of the eastern margin of The Oiga Graben. Under the influence of geological structure, the folds, joints and fractures in the tunnel area are extremely developed, with poor interlayer bonding and separation, and the strength of the structural plane is in the residual state. The unconformity of the Angle between rock stratum makes the integrity of rock mass seriously damaged and causes the rock mass structure to be loose and broken.
In addition, the rock mass is affected by multiple magmatic intrusions, and the granite in the large deformation section shows obvious alteration characteristics. The rock mass is a cataclastic dispersion structure with low strength, and the rock block is easily broken by hand. The whole self-stabilization ability of surrounding rock is low, and the surrounding rock is easy to be deformed after excavation.
(2) High ground stress
Through the ground stress tests, within the test range of 604.45m, by hydraulic fracturing technique, the maximum horizontal ground stress of the rock mass is 17.72 MPa, with a maximum ratio of 1.76 to the vertical stress. There is large structural stress in the rock mass with a dominant direction of N9°W–N7°E. Due to the low strength of altered granite in the large deformation section of Zangga tunnel, the surrounding rock is prone to deformation under the influence of high ground stress.

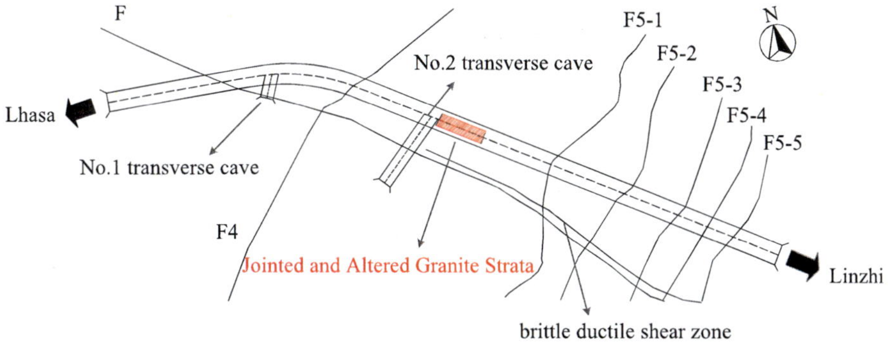


**Note**: F—Lalong-Zangga fault zone; F4—Zhengga fault zone ; F5—Oiga-Luobusha fault zone.

**Fig. 9.** Fault zone distribution diagram of Zangga tunnel

(3) Other factors
Scholars find that the shape of the section plays a certain role in controlling the convergent deformation of surrounding rocks. The closer the cross-section shape is to the circle, the stronger the resistance to tunnel convergent deformation is. However, Zangga tunnel is a single-line railway tunnel, which high-span ratio of the excavated section is relatively large (about 1.22) and the curvature of the sidewall is small. The shape of this section has no obvious advantage in the control of convergent deformation. Therefore, under the condition of unfavorable section shape, large tunnel convergent deformation is easy to occur after tunnel excavation due to the influence of high tectonic stress. In addition, groundwater, bench length and impact of construction are also factors that cause large deformation.

# 4 Control Measures for Large Deformation of Surrounding Rock

Through the above analysis, the deformation characteristics of surrounding rock and the influencing factors of the jointed and altered granite stratum of Zangga tunnel section are clarified. Based on the concept of "combination of release and resistance" control measures for large deformation of soft rock tunnel, the comprehensive control measures for surrounding rock deformation are put forward, including: Adjusting the curvature of sidewall; Adopting advance reinforcement and double-layer primary support; Increasing the reserved deformation; Setting up long anchor rod. Additional auxiliary construction measures such as temporary invert and anchor pipe are also very important control measures. The field test section of surrounding rock is given priority to with V level of surrounding rock, joint and fissure development. The rock mass is of fragmentation structure with low strength and easy to disintegrate in water. There is no significant difference between DK169+025–DK169+140.

### 4.1 Control Measures of the Field Test Section

(1) Support method

Compare with $V_c$ composite lining structure, $III_B$ lining structure (shown in Fig. 10), applied in the field test section, adjusts the sidewall curvature (lateral wall arch camber increased 500 mm) and adopts a more powerful supporting measures in advance reinforcement. The primary support is strengthened from one layer to the double layer, which improves the stiffness of the primary support. The reserved deformation value of primary support is increased. The reserved deformation amount of the first layer is 400 mm (The value at the invert is 0 mm), and that of the second layer is 200 mm. Moreover, $III_B$ lining structure strengthened the anchor design parameter, also added some auxiliary construction measures such as the temporary inverted arch and lock foot anchor pipe. Support parameters of the field test section are shown in Table 1.

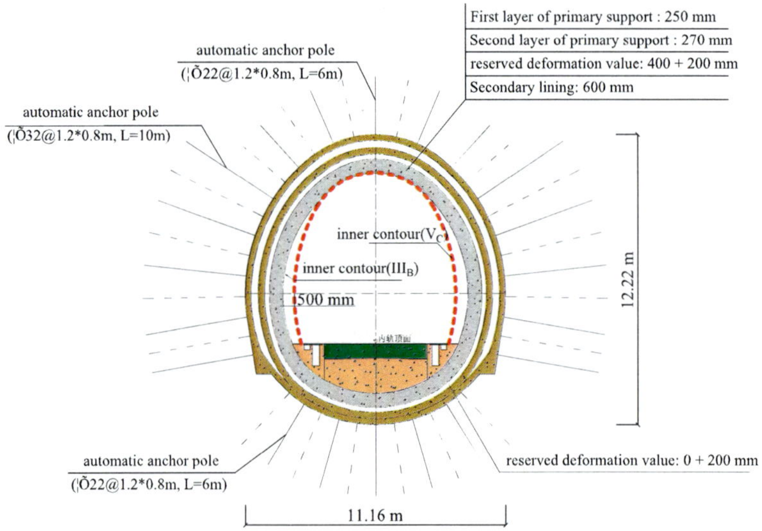


**Fig. 10.** Section diagram of lining structure of type $III_B$ test section

(2) Excavation method

The field test section is constructed by three-bench method, and the length of the bench is strictly controlled. Specific construction steps are as follows: ① The advanced reinforcement of the arch of the tunnel face. ② Upper bench excavation, followed by the first layer of primary support. The excavation height of the upper bench is 4.4 m. ③ Excavate and support on the right side of the middle bench, with the excavation height of 1.8 m and 5 m from the upper bench; The left side of the middle bench shall be excavated and supported, 3 m from the right side. ④ Excavate and support on the right side of the bottom bench, and the excavation height is 4.2 m, 3 m from the left side of the middle bench; Excavation and support shall be carried out on the left side of the bottom bench, 3 m from the right side. ⑤ In

**Table 1.** Support parameters of the field test section.

| Mileage | Advanced reinforcement measures | Support measures |
|---|---|---|
| DK169+140–DK169+200 | Pipe shed (Φ89 mm@6.6 * 0.4 m, L = 10 m; 29 rings per ring) Or small pipe (Φ42 mm@1.8 * 0.4 m, L = 3.5 m, 29 rings per ring) | It is shown in Fig. 10 |
| DK169+200–DK169+263 | Except small pipe (Φ42 mm@2.4 * 0.4 m, L = 3.5 m, 29 rings per ring), the other advance support parameters are the same as the above contents | Based on the design parameters in Fig. 10, the following contents are added: the I20b temporary invert with a distance of 1.2 m at the upper bench, and the four anchor pipes with a length of 4 m and a diameter of 42 mm on the steel frame at the arch foot on the two sides of the middle bench |

time use the self-feeding anchor (Φ32). After the primary support on the left and right sides of the bottom bench is completed, the second layer of primary support is applied on the basis of the first layer of primary support at the arch and the sidewall. ⑥ According to the monitoring and measurement results, determine the best time for the secondary lining and apply the secondary lining. Construction sequence of the three-bench method is shown in Fig. 11.

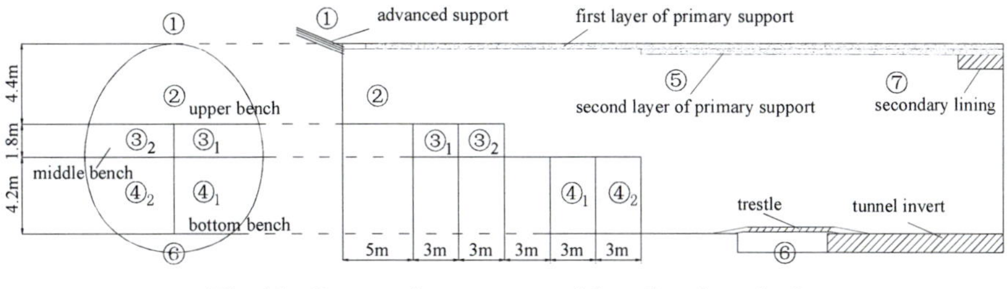


**Fig. 11.** Construction sequence of three-bench method

## 4.2 Deformation of Surrounding Rock in the Field Test Section

After adopting control measures such as double-layer support, the cumulative deformation curve of surrounding rock in the field test section DK169+140–DK169+263 is statistically obtained, as shown in Fig. 12. The statistical data of surrounding rock deformation in large deformation section and test section are shown in Table 2.

It can be seen from the figure and the table that the deformation of surrounding rocks in the test section is well controlled. The maximum crown settlement in the test

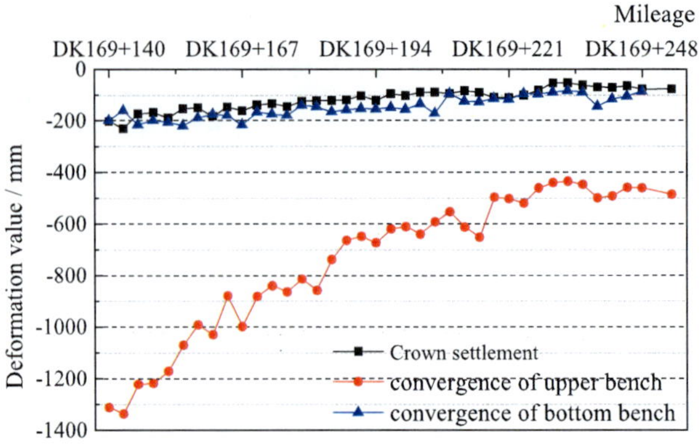


**Fig. 12.** Cumulative deformation of surrounding rock at the section of DK169+140–DK169+263

**Table 2.** Statistical data of large deformation section and test section

| Mileage | Crown settlement/mm | | Upper bench convergence/mm | | Bottom bench convergence/mm | | Reserved deformation value/mm | Deformation duration/d |
|---|---|---|---|---|---|---|---|---|
| | Average | Maximum | Average | Maximum | Average | Maximum | | |
| DK169+025–DK169+140 | 248.3 | 656 | 942.7 | 3 023.1 | | | 30–100 | No convergence trend (within 45d) |
| DK169+140–DK169+200 | 146.5 | 231.1 | 927.9 | 1 338.4 | 172.6 | 220.8 | 400+200 | 60–75 |
| DK169+200–DK169+263 | 82.09 | 110.4 | 521.52 | 653.5 | 114.63 | 144.5 | 400+200 | 35–50 |

section is 231.1 mm and located in section DK169+143. The maximum value of tunnel convergence is 1338.4 mm, located at the upper bench on DK169+143 section, and the overall deformation value of the field test section is less than that of the large deformation section. The control effect can be reflected from the following aspects.

(1) The deformation value of surrounding rock is controlled. After adjusting the curvature of the sidewall and adopting the double-layer primary support structure, the surrounding rock deformation value of the test section is lower than that of the large deformation section, and the deformation value is basically within the safe range.
(2) Deformation rate of surrounding rock is significantly reduced. In the 3–5 days after the excavation of the upper bench, the average deformation rate of the upper bench convergence in the surrounding rock of the large deformation section is over 50 mm/d, while the corresponding deformation rates in the test section DK169+140–DK169+200 and DK169+200–DK169+263 are within 30–50 mm/d and 10–30 mm/d respectively. The deformation rate is well controlled.

(3) The deformation duration of surrounding rock is shortened. The deformation of the surrounding rock in the large deformation section did not show a convergence trend within about 45 days, while the deformation of the surrounding rock in the test section was stable within about 35–75 days, and the deformation duration was significantly shortened.
(4) The control effect of DK169+200–DK169+263 section is better than that of DK169+140–DK169+200 section. Through the comparison of deformation data, it can be seen that the deformation duration of section DK169+200–DK169+263 is relatively short after the addition of temporary invert and anchor pipe. At the same time, the average cumulative convergent deformation rate of the upper bench in each construction stage of this section is only 48.72%, 57.14% and 77.54% of the DK169+140–DK169+200 sections, making its cumulative deformation value small. The additional auxiliary measures such as the temporary invert and anchor pipe on the upper bench have a positive effect on the deformation control of the formation.

### 4.3 Application Effect

The large deformation section of DK169+025–DK169+140 started construction on April 7, 2017, and completed invert excavation on January 13, 2018 (281d). Since most of the primary support deformation exceeded the limit value, the construction unit removed and replaced it, which consumed a total of 403 days. The excavation of the test section DK169+140–DK169+263 takes a total of 217 days, and the total construction time of the whole section is 248 days. The comparison shows that the tunnel construction efficiency is obviously improved.

At present, Zangga tunnel has been successfully completed. After the comprehensive control measures of surrounding rock deformation have been adopted, no abnormal phenomena have appeared in the construction process of the test section and subsequent such stratum, and the construction effect is good. It can be seen that the comprehensive control measures of surrounding rock deformation, which mainly include adjusting the curvature of sidewall, the combination of long and short bolts and double-layer primary support, have a good effect on the deformation control of surrounding rock.

## 5 Conclusions

When Zangga tunnel passed through the jointed and altered granite stratum, the surrounding rock appeared large deformation phenomenon, and the primary support structure was seriously damaged. The support design has insufficient adaptability to the stratum. Based on the above deformation results of the surrounding rock, this paper analyzed the deformation characteristics of the surrounding rock in the jointed and altered granite stratum, discussed the influencing factors of the surrounding rock deformation, and studied the deformation control measures through field tests. The main conclusions are as follows:

(1) During the construction of the jointed and altered granite stratum with high ground stress tunnel, the surrounding rocks show the characteristics of large magnitude

and fast speed, and the tunnel convergent deformation is greater than the crown settlement. The deformation of surrounding rock is divided into acceleration and development stages. The deformation rate does not decrease significantly after the enclosing structure is closed and formed, and the deformation of surrounding rock continues to increase. The original design cannot control the deformation.

(2) The large deformation of surrounding rock is caused by geological reasons, high ground stress and other reasons. Among them, the influence of active fault on surrounding rock and alteration of surrounding rock are the internal causes of large deformation of tunnel, while high ground stress is the external factor. In addition, the influence of section shape, excessively long construction bench and construction on surrounding rock deformation cannot be ignored.

(3) Combined with the characteristics and influencing factors of surrounding rock deformation, the comprehensive control measures of surrounding rock deformation are put forward, including adjusting the curvature of surrounding rock, combining long and short bolt and double-layer main support. The field application results show that the deformation value, rate and duration of surrounding rock can be controlled, the construction effect is good, and the construction efficiency is greatly improved. The research can provide reference for subsequent construction and similar projects.

**Acknowledgements.** This study was supported by the National Natural Science Foundation of China (U1934211).

## References

Coggan, J.S., Chilton, J.L., Stead, D., et al.: Effects of alteration on the engineering behaviour and intact rock fracture characteristics of granite under uniaxial compression. In: Rock Mechanics: Meeting Society's Challenges and Demands (2007)

Khanlari, G.R., Naseri, F.: Investigation of physical deterioration of Malayer granitic rocks using a new weathering coefficient ($Kr_4$). Environmental Earth Sciences **75**(5), 1–14 (2016). https://doi.org/10.1007/s12665-015-5046-7. Article number: 414

Meller, C., Kohl, T.: The significance of hydrothermal alteration zones for the mechanical behavior of a geothermal reservoir. Geoth. Energy **2**(12), 1–21 (2014). https://doi.org/10.1186/s40517-014-0012-2. Article number: 12

Meng, L.B., Li, T.B., Jiang, Y., et al.: Characteristics and mechanisms of large deformation in the Zhegu mountain tunnel on the Sichuan-Tibet highway. Tunn. Undergr. Space Technol. **37**, 157–164 (2013)

Tao, Z.G., Cao, J.D., Yang, L., et al.: Study on deformation mechanism and support measures of soft surrounding rock in Muzhailing deep tunnel. Adv. Civ. Eng. (2), 1–14 (2020)

Xiao, X.W., Wang, L.C., Yang, J.S., et al.: Cause analysis and treatment scheme for bottom heave of Ballastless track tunnel in nearly horizontally Interbedded Rock Mass with high geostress. China Railway Sci. **37**(1), 78–84 (2016). (in Chinese)

# Multi-channel Singular Spectrum Analysis of Geocenter Motion and Precise Prediction of Earth's Motion

Xin Jin[1], Xin Liu[1(✉)], Jinyun Guo[1], and Yi Shen[2]

[1] College of Geodesy and Geomatics, Shandong University of Science and Technology, Qingdao 266590, China
skd994268@sdust.edu.cn

[2] School of Geographic Sciences, Xinyang Normal University, Xinyang 464000, China

**Abstract.** Geocenter is the center of the mass of the Earth system including the solid Earth, ocean and atmosphere. The time-varying characteristics of geocenter motion (GCM) can reflect the redistribution of the Earth's mass and the interaction between solid Earth and mass loading. The GCM also provides important support for millimeter level, dynamic real-time global reference framework. The GCM products determined from satellite laser ranging data released by the Center for Space Research through Jan. 1993 to Feb. 2017 is utilized to determine the periods and the long-term trend of GCM by using multi-channel singular spectrum analysis (MSSA). The results show that the GCM has the amplitude between 1 mm and 10 mm, and has obvious seasonal characteristics of the annual, semiannual, quasi-0.6 years and quasi-1.5 years in X, Y and Z directions, the annual characteristics make great domination. It also shows long period terms of 6.09 years as well as the trends of 0.05 mm/yr, 0.04 mm/yr and −0.10 mm/yr in the three directions, respectively. The MSSA is applied to predict GCM and combines it with the linear model (LM) and autoregressive moving average model (ARMA) to predict GCM with ahead 2 years. The results show that the LM+MSSA+ARMA model can effectively predict GCM parameters, and can provide prediction precision of 1.5 mm, 1.1 mm and 3.5 mm in X, Y and Z directions, respectively.

**Keywords:** Multi-channel singular spectrum analysis · Geocenter motion · Prediction · ARMA

## 1 Introduction

The center of mass (CM) of the Earth is defined as the center of the mass of the entire Earth including solid Earth, ocean and atmosphere (Montag 1999; Pavlis 1999) (see Table 1 for a list of abbreviations used in the paper). Sea level change, glacier melting, atmospheric circulation, and mantle convection result in the CM movement relative to the center of the figure (CF) of the Earth surface, which reflects the global mass redistribution and the interaction between solid Earth and mass loading (Wu et al. 2012; Cheng et al. 2013a). The geocenter motion (GCM) is helpful to study the problem that


H. Khabbaz et al. (Eds.): GeoChina 2021, SUCI, pp. 29–45, 2021.
https://doi.org/10.1007/978-3-030-79650-1_3

the International Terrestrial Reference Frame (ITRF) implementation is not completely consistent with the International Terrestrial Reference System (ITRS) definition, which will further improve the precision of the reference frame origin (Blewitt 2003; Dong et al. 2003; Metivier et al. 2010; Altamimi et al. 2011). The GCM has an important impact on the reference frame transformation of Satellite Laser Ranging (SLR), Global Navigation Satellite System (GNSS) and Doppler Orbitography and Radio-positioning Integrated by Satellite (DORIS) system (Wei et al. 2016; Riddell et al. 2017). It is also an important topic for studying the Earth's mass redistribution, such as ocean tide, glacial isostatic adjustment, atmospheric and ocean circulation, geodynamic process in the Earth's interior (Trupin et al. 1992; Dong et al. 1997; Watkins and Eanes 1997).

**Table 1.** Abbreviations used in this paper

| Acronym | Full name | Acronym | Full name |
|---|---|---|---|
| CM | Center of the mass | ITRS | International Terrestrial Reference System |
| CF | Center of the figure | SLR | Satellite Laser Ranging |
| GCM | Geocenter motion | GNSS | Global Navigation Satellite System |
| CSR | Center for Space Research | DORIS | Doppler Orbitography and Radio-positioning Integrated by Satellite |
| MSSA | Multi-channel singular spectrum analysis | IGS | International GNSS Service |
| SSA | Singular spectrum analysis | GRACE | Gravity Recovery and Climate Experiment |
| LM | Linear model | UTOPIA | University of Texas Orbit Determination Program |
| ARMA | Autoregressive moving average model | LLISS | Large Linear System Solver |
| ITRF | International Terrestrial Reference Frame | | |

SLR data were processed to estimate effectively GCM (Bouille et al. 2000; Crétaux et al. 2002). The fewer SLR sites, the maldistribution of sites and the absence of SLR data on the ocean lead to the deterioration of the measurement precision. GNSS data were also used to determine GCM (Blewitt and Clarke 2003). Although there are more International GNSS Service (IGS) stations all over the world, GNSS satellite orbital altitude is so high that the precision of GNSS-derived GCM is lower than that of SLR-derived products. The DORIS-derived GCM in Z direction is very different from the other two directions and its precision is only up to centimeter-level. DORIS-derived GCM has the lowest precision among the series estimated by these three space geodetic techniques (Kuzin and Tatevian 2005; Kong et al. 2017). The time series of GCM in different time-span can be accurately estimated by using the SLR data of Lageos-1, Lageos-2 and other geodynamical satellites.

The wavelet transformation, least squares spectral analysis, singular spectrum analysis (SSA) can be used to discover the characteristics of GCM in X, Y and Z directions (Guo et al. 2009; Wei et al. 2016). However, when analyzing the GCM, these methods cannot take into account the correlation of all directions of GCM. Multi-channel singular spectrum analysis (MSSA), as the extended form of SSA, is one of the effective statistical data analysis methods in oceanography, geoscience, meteorology and other fields (Vautard and Ghil 1989; Wang et al. 2016; Zotov et al. 2016; Zhou et al. 2018). The MSSA is a method for analyzing nonlinear time series. It is also able to denoise data, extract periodic oscillation signals, identify trends from multidimensional time series and build prediction models (Shen et al. 2017, 2018). Compared with SSA, during the process of multidimensional time series, the correlations among different channels are taken into account, so we apply the MSSA method to the analysis of GCM and study the ability to extract periodic signals of GCM.

The monitoring and modeling of GCM is a key issue for constructing a millimeter-level, dynamic and real-time global reference frame. However, due to the complexity of obtaining multi-source observations and data processing, The GCM parameters cannot be obtained in real-time or quasi-real-time (Altamimi et al. 2011; Zhao et al. 2019). Therefore, the MSSA model is applied to predict GCM parameters and proposes a GCM prediction method that combines linear model (LM), MSSA and autoregressive moving average model (ARMA).

SLR-derived GCM series from Jan. 1993 to Feb. 2017 updated by the Center for Space Research (CSR), the Texas University at Austin, are used to study GCM variation in this paper. The trend and periodic variations of GCM are investigated by using MSSA. Finally, based on historical GCM data, the fusion method of LM, MSSA and ARMA models is used to predict GCM parameters.

## 2 Data Collection and Methodology

### 2.1 SLR-Derived GCM Products

The GCM products used in this paper are obtained from CSR at University of Texas website (http://ftp.csr.utexas.edu/pub/slr/geocenter/). The GCM products (GCN_L1_L2_30d_CF-CM) were solved with UTOPIA and LLISS from SLR data of geodynamical satellites (e.g. Lageos-1/2, Starlette, Ajisai and Stella) in SLRF2014 (Cheng et al. 2013a; Pearlman et al. 2019). CF-CM is intended to reflect the true degree-1 mass variations without being affected by the higher-degree site loading effects (particularly at the annual frequency) (Ries 2016). The GCM products are often used to study the local and global mass balance with GRACE and are currently the best geocenter coordinate result recognized internationally. Here we download the GCM products from Jan. 1993 to Feb. 2017.

### 2.2 Multi-channel Singular Spectrum Analysis

There is a time series $x_{li}(l = 1, ..., L; i = 1, ..., N)$ in which $l$ is channel number and $i$ is time sequence number. The rank of $x_{li}$ is arranged according to the time delay phase space $M$ $(1 \le M \le N/2)$ that is the window length and also called the step number of time lag. The integer multiple of the main cycle is generally chosen as one window length in MSSA (Oropeza and Sacchi 2011; Shen et al. 2018).

The trajectory matrix of the channel $l$ is

$$\mathbf{X}_i = \begin{bmatrix} x_{1i} & x_{2i} & \cdots & x_{Ki} \\ x_{2i} & x_{3i} & \cdots & x_{K+1i} \\ \vdots & \vdots & \vdots & \vdots \\ x_{Mi} & x_{M+1i} & \cdots & x_{Li} \end{bmatrix} (1 \le i \le N) \quad (1)$$

where $K = N - M + 1$. The multi-channel trajectory matrix can be indicated as

$$\mathbf{X} = [\mathbf{X_1}, \mathbf{X_2}, \cdots, \mathbf{X_N}]^{\mathbf{T}} \quad (2)$$

Matrix $\mathbf{X}$ has $L \times M$ rows and $N - M + 1$ columns. Similar to SSA, the next step is to decompose the singular value of $\mathbf{X}$. We define the matrix $\mathbf{S} = \mathbf{X}\mathbf{X}^T$, where $\mathbf{X}^T$ is the transposed matrix of $\mathbf{X}$. Suppose that $\lambda_1, \cdots, \lambda_M$ are the eigenvalues of matrix $\mathbf{S}$, that is, the singular values. These eigenvalues are arranged in the descending order. The larger singular value generally represents the larger energy signal and the smaller one corresponds to the noise part. Matrix $\mathbf{X}$ can be expressed in the elementary matrix as

$$\mathbf{X} = \mathbf{P_1} + \mathbf{P_2} + \cdots + \mathbf{P_D} \quad (3)$$

where $D$ represents the number of singular values, and $\mathbf{P}_i = \mathbf{S}_i\mathbf{U}_i\mathbf{V}_i^T$ in which $\mathbf{U}_i$ is the temporal empirical orthogonal function and $\mathbf{V}_i$ is the temporal principal components.

The GCM time series contain different signals, such as annual term and semi-annual term. It is necessary to use the w-correlation method (Hassani 2007) to merge elementary matrix $\mathbf{P}_i$ representing the same signal into a group. Suppose that the time series after reconstruction of each elementary matrix $\mathbf{P}_i$ is $\mathbf{Y}_i$, the correlation of any two reconstructed time series can be expressed by w-correlation as

$$\rho_{i,j}^{w} = \frac{(Y^{(i)}, Y^{(j)})}{\|Y^i\|_w \|Y^j\|_w}, (1 \le i, j \le L) \quad (4)$$

Where $\|Y^i\|_w = \sqrt{(Y^{(i)}, Y^{(i)})}$, $(Y^{(i)}, Y^{(j)}) = \sum_{k=1}^{N} w_k y_k^i y_k^j$, and $w_k = \min(k, M, N - k)$. The larger the absolute value of $\rho_{i,j}^{w}$ is, the greater the correlation of the corresponding components of $i$ and $j$ is, which should be classified as the same periodic signal component. Then the corresponding trajectory matrix has been build.

The corresponding group of trajectory matrix is converted into a new time series with the length of $N$, which is called the reconstructed component (Golyandina et al. 2001). Then the reconstructed component (RC) is

$$x_{li}^{k} = \begin{cases} \frac{1}{i}\sum_{j=1}^{i} a_{i-j}^{k} E_{lj}^{k} & 1 \le i \le M-1 \\ \frac{1}{M}\sum_{j=1}^{M} a_{i-j}^{k} E_{lj}^{k} & M \le i \le N-M+1 \\ \frac{1}{N-i+1}\sum_{j=i-N+M}^{M} a_{i-j}^{k} E_{lj}^{k} & N-M+2 < i \le N \end{cases} \tag{5}$$

# 3 Analysis of GCM

## 3.1 GCM Seasonal Variation

The GCM time series comprising monthly data ($L = 290$) from Jan. 1993 to Feb. 2017 are used to analyze GCM variation. The trajectory matrix of GCM are decomposed by selecting the window $M = 108$ (month), which is determined by the period of the annual term of GCM and many practical experiments. Singular values in descending order are shown in Fig. 1.

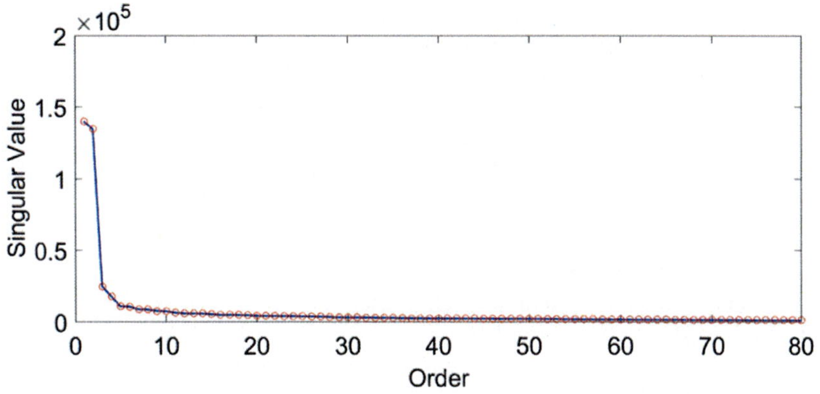


**Fig. 1.** Singular values of GCM series determined by MSSA

As shown in Fig. 1 that the values starting from the 20th singular value are already smaller than the first 4th previous singular values, and the values after 20th order change smoothly so that they can be ignored. The w-correlations $\rho_{i,j}^{w}$ of the first 20 reconstruction parts of GCM in the three directions except for the trend term are shownin Fig. 2.

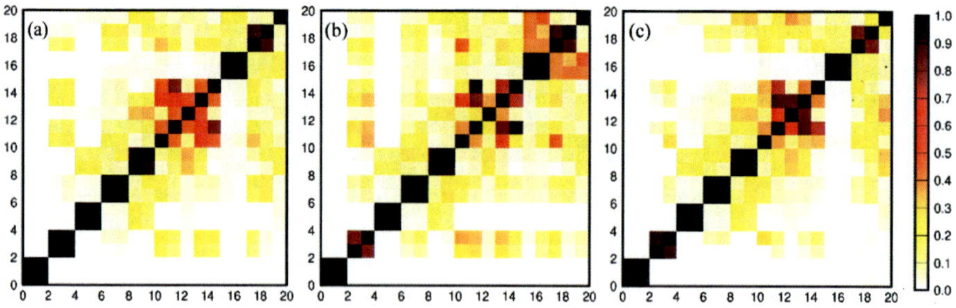


**Fig. 2.** W-correlations for the first 20 reconstructions. (a) X direction, (b) Y direction, and (c) Z direction.

The greater w-correlation $\rho_{i,j}^{w}$ means the corresponding components belong to the same periodic term. As shown in Fig. 2 that the time series in the X direction is not completely separated from each other when $i > 10$ and the separation effect in both Y and Z directions also deteriorates after $i > 10$, which may be caused by noises. Hence, the first 10 groups (RC1, RC2, ..., and RC10) are used to reconstruct the new GCM series. RC1 and RC2 represent a periodic term in the original series; RC3 and RC4 represent the other one; RC5 and RC6, RC7 and RC8, RC9 and RC10 can be combined into one periodic component, respectively. Figure 3 shows the original GCM series in X, Y and Z directions except for the trend term and the time series reconstructed by the first 10-order reconstruction components.

As shown in Fig. 3 that the fluctuation ranges of raw GCM in X and Y directions are smaller than that in Z direction. Although the offset of CM relative to CF in the Z direction is large, the fluctuation amplitude is small, and most of them are negative.

The correlation coefficients between the time series reconstructed by the first 10 singular vectors and the corresponding original time series in X, Y and Z directions are 73.34%, 86.78% and 83.84%, respectively, which indicates that they have good consistency. It shows that MSSA model can effectively extract relatively complete information about the main components in the three directions.

Table 2 shows the singular spectrum values and the variance contributions of GCM time series calculated by MSSA from 1993 to Feb. 2017. The variance contribution of the first 10-order reconstruction components has reached 65.80%, which can characterize the variation of GCM effectively. Furthermore, the variance contribution of reconstruction RC1 and RC2 is significantly larger than that of other components, which indicates that the corresponding cyclophysis is most obvious. The first 10-order reconstruction components can be combined into five periodic terms according to the principle of w-correlation. Figures 4, 5 and 6 show the reconstructed component and the corresponding power spectrum in the three directions.

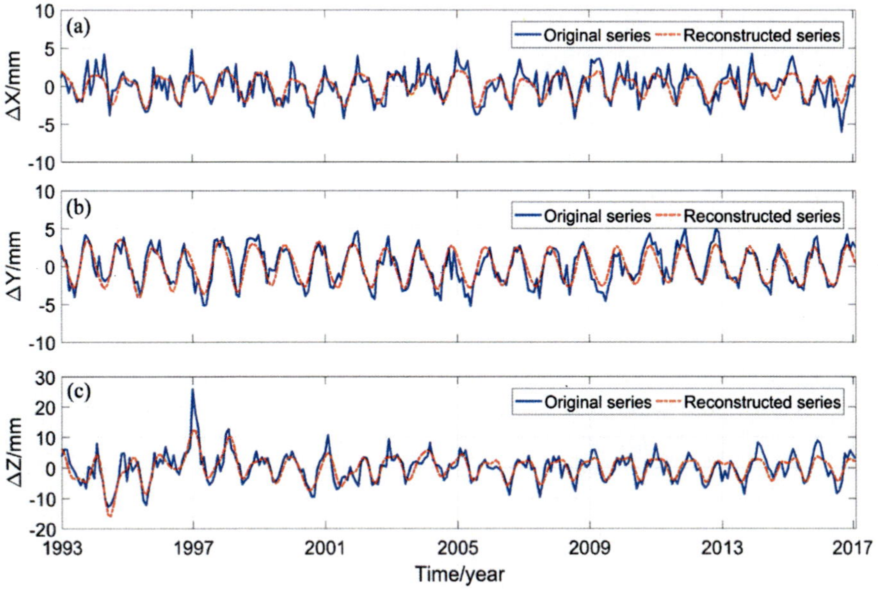


**Fig. 3.** Time series of GCM from 1993 to 2017.2. (a) X direction, (b) Y direction, and (c) Z direction.

**Table 2.** Singular values and variance contributions of the first 10 orders based on MSSA

| Order/RC | Singular spectrum value | Variance contribution |
|---|---|---|
| 1 | 139651.7 | 24.90% |
| 2 | 134495.7 | 23.98% |
| 3 | 24465 | 4.36% |
| 4 | 17625.2 | 3.14% |
| 5 | 10608.3 | 1.89% |
| 6 | 10477.4 | 1.87% |
| 7 | 8517.8 | 1.52% |
| 8 | 8476.8 | 1.51% |
| 9 | 7423.4 | 1.32% |
| 10 | 7285.8 | 1.30% |

As shown in Figs. 4, 5 and 6 that the five main periodic terms of GCM in three directions are basically the same. There is the annual term, semi-annual term, quasi-0.6 years term, quasi-1.5 years term and long-term term. The cycles are respectively 0.99 years, 0.5 years, 0.58 years, 1.47 years and 6.09 years. The fifth main periodic term is also mixed with a small periodic term of about 1.4 years. The variance contributions

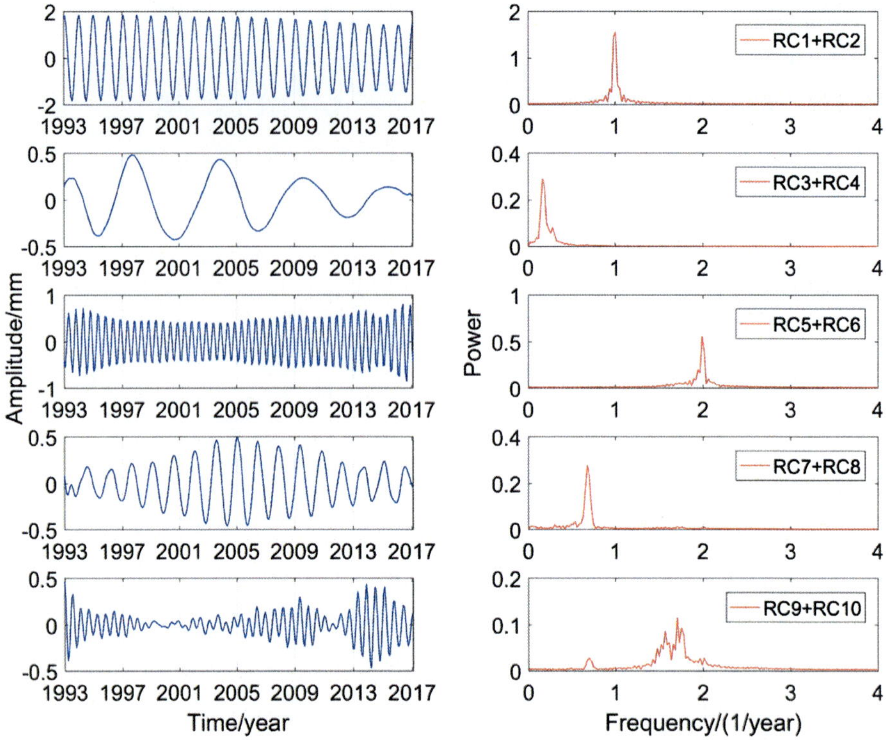


**Fig. 4.** Combination of RC and the power spectrum analysis in X direction

of these five main periodic components in the three directions are different, and annual characteristics make great domination.

The annual periodic oscillations in the three directions are relatively stable, and the periodicity is the most obvious, with amplitudes of about 1.7 mm, 2.8 mm and 4.4 mm respectively. The valley value of the annual variation of GCM in X direction occurs from August to September and the peak value appears from February to March. The valley value of the annual variation of GCM in Y direction occurs from May to June and the peak value appears from November to December. The valley value of the annual variation of GCM in Z direction occurs from July to August and the peak value appears from January to February. The semi-annual term oscillation is not relatively obvious. Although the semi-annual variations of 0.5 years are shown in these three directions, the corresponding amplitude variation characteristics are not the same. In the past decade, the amplitude of the semi-annual cycle in three directions have gradually increased. The cyclophysis of quasi-1.5 years and quasi-0.6 years, one of the five main periods of GCM, also exhibits strong seasonal characteristics. The oscillation characteristics of quasi-1.5 years and quasi-0.6 years period in Y and Z direction are similar, and the amplitude gradually decreases with time. The oscillation of quasi-1.5 years period in X direction has obvious fluctuations and reached its maximum in February 2005. The

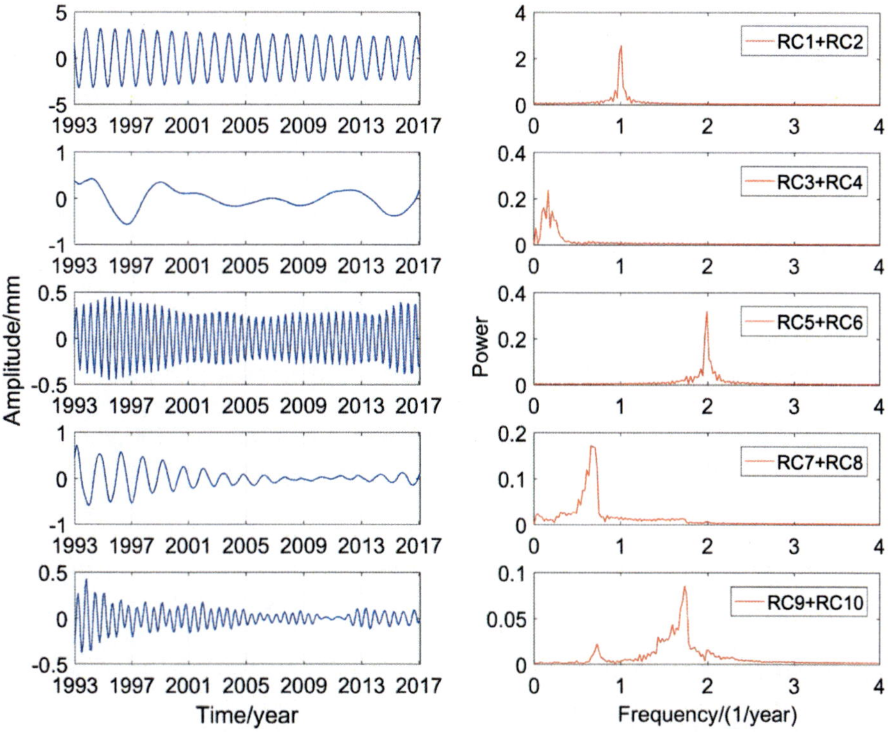


**Fig. 5.** Combination of RC and the power spectrum analysis in Y direction

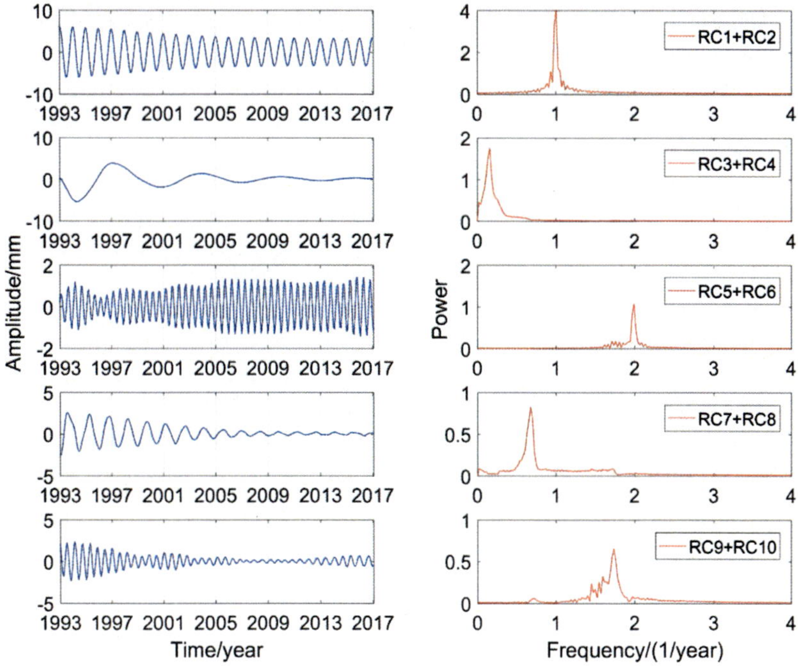


**Fig. 6.** Combination of RC and the power spectrum analysis in Z direction

seasonal characteristic of the quasi-0.6 years period in X direction is obvious, which may be due to the inclusion of many other signals in the period.

The annual term, semi-annual term, quasi-0.6 years term and quasi-1.5 years term mentioned above belong to the seasonal cycle of GCM. The solar radiation, changes in the gravitational field and other Earth external energy cause the surface mass redistribution of land water, ocean and atmosphere, which results in the significant seasonal GCM. The major reason for the seasonal cycle of GCM is the seasonal variation of land water storage (Guo et al. 2008; Feissel-Vernier et al. 2006). The exchange of water mass in the Earth's hemisphere has a clear annual cycle. Greater water mass in the northern hemisphere appears during June-August, instead of appears during December-February in the southern hemisphere (Blewitt et al. 2001; Baur et al. 2013). The peak and valley values of the annual term in Z direction may be the reflection of water mass redistribution.

The long period terms in X, Y and Z directions analyzed by MSSA are all 6.09 years. The major reason for the secular variation of center of mass of the Earth system is the glacial isostatic adjustment. The influence of the glacial ablation on the solid Earth causes GCM velocity of less than 1 mm/yr (Klemann and Martinec 2011). The amplitude of the long-period term in the X direction is maintained within 0.5 mm, but the amplitude in the Y and Z directions has a sudden increase of 1 mm in 1997, which may be caused by the El Ninõ that was strongest in the 20th century (Guo et al. 2009).

### 3.2 GCM Trend Variation

The GCM series in X, Y and Z directions are directly decomposed and analyzed by using MSSA to determine their trends. The GCM trend variation from 1993 to 2017.2 is shown in Fig. 7. The secular velocity of GCM is calculated by least square method. Table 3 shows the comparisons among the proposed method and the reported methods.

As shown in Fig. 7 that the trend variation in the three directions of GCM are nonlinear. After 1997, the moving direction of GCM in the three directions remained stable. It can be seen from Table 3 that the long-term speed of GCM obtained from different data and time has a relatively large difference, and there is no accurate reference standard at present. In this paper, the variation rates in X, Y and Z are 0.05 mm/yr, 0.04 mm/yr and −0.10 mm/yr, respectively, all of which are less than 1 mm/yr. After 1997, the long-term variation rate of GCM in Z direction is negative, which may be caused by glacial isostatic adjustment (Wu et al. 2010).

Due to the local expansion of the zero-degree hemisphere and the condensation of the 180-degree hemisphere, the variation rate of GCM in X direction is positive, which shows that CM moves towards the X direction relative to CF. Besides, the maximum variation of atmospheric pressure occurs in Central Asia (Blewitt et al. 2001; Greff-Lefftz and Legros 2007). The solid Earth is a viscous elastomer so that the maximum fluctuation of atmospheric pressure in Central Asia may have a certain impact on GCM in Y direction, which results in the positive of the secular velocity.

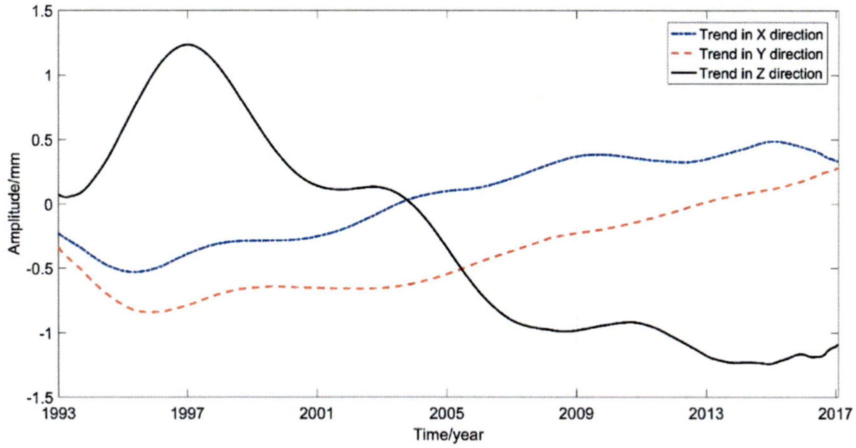


**Fig. 7.** GCM trend variation by using MSSA

**Table 3.** Secular velocity of GCM in different studies (mm/yr)

| Literature | Data and model | X | Y | Z | Time span |
|---|---|---|---|---|---|
| This paper | SLR | 0.05 ± 0.003 | 0.04 ± 0.004 | −0.10 ± 0.01 | 1993–2017.2 |
| Guo et al. (2009) | SLR | −0.26 ± 0.02 | 0.43 ± 0.02 | 0.50 ± 0.02 | 1993–2006 |
| Kuzin et al. (2010) | DORIS/INA | −1.19 ± 0.07 | −0.12 ± 0.07 | −0.28 ± 0.31 | 1993–2009 |
| Rietbroek et al. (2012a) | GRACE/Jason-1/GIA | −0.28 | 0.43 | −1.08 | 2003–2008 |
| Sun et al. (2016) | GRACE/OMCT/ICE-5G_VM2 | −0.03 ± 0.03 | 0.11 ± 0.02 | −0.21 ± 0.04 | 2002.6–2014.5 |
| Sun et al. (2016) | GRACE/OMCT/ICE-6G_VM5a | −0.06 ± 0.03 | 0.07 ± 0.02 | −0.33 ± 0.04 | 2002.6–2014.5 |

# 4 Prediction of GCM

## 4.1 Principle of Prediction

With the continuous development and improvement of space technology, there is a growing demand for the prediction parameters of GCM, such as millimeter level, dynamic real-time global reference framework and high-precision satellite navigation and positioning (Metivier et al. 2010; Zhao et al. 2019). To meet this demand, therefore, the combined method of LM and MSSA and ARMA is applied to predict the GCM parameters. The flow chart of prediction is shown in Fig. 8.

(1) First, perform linear fitting on the GCM series, establish a linear model and make predictions. Assuming that the GCM time series is $X_t$, the least squares method is used to linearly fit it, and the model is expressed as:

$$X_t = \beta_0 + \beta_1 \mathrm{t} \tag{6}$$

Where $\beta_0$ is a constant, and $\beta_1$ represents the linear trend change rate.

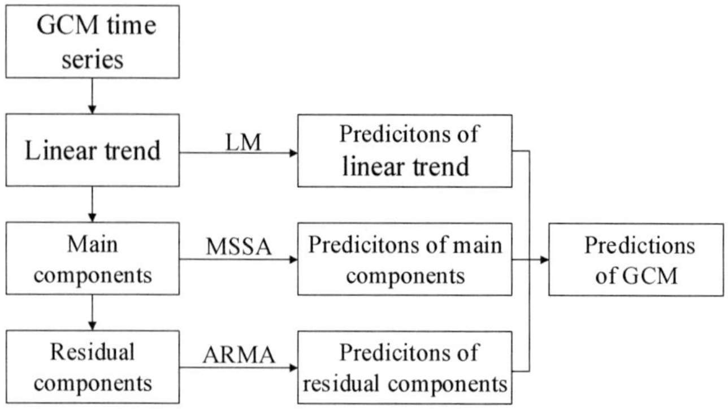


**Fig. 8.** The prediction method of GCM

(2) MSSA is used to decompose the GCM series of the de-linear trend, and the appropriate main component terms are selected for the main component prediction of the GCM. The main component prediction method of MSSA is as follows:

- The number of predictions of GCM is $N$, assuming that the time series of GCM without linear trend is $Y_t$, $N$ zeros are added at the end of $Y_t$ to form a new prediction sequence $Y_{t+N}$;
- The new prediction sequence $Y_{t+N}$ is decomposed by MSSA, and the $N$ values at the end of the first RC (RC1) are used to replace the corresponding prediction values of the new sequence. This process is repeated until the RMS value of the two replacements data is less than 0.001 mas;
- RC2 is added to reconstruct the prediction data, that is, the prediction data is obtained by linear superposition of RC1 and RC2. Step 2 is repeated until RC1…RCi is linearly added to the prediction data, and the predictions using MSSA can be obtained.

(3) ARMA is used to model and forecast the residual components reconstructed by MSSA. Assuming that the sequence of residual items is $Z_t$, the ARMA model is expressed as:

$$Z_t = c + \sum_{i=1}^{p} \phi_i Z_{t-i} + \varepsilon_t + \sum_{j=1}^{q} \theta_j \varepsilon_{t-j} \tag{7}$$

Where $p$ represents the order of the AR model, $q$ represents the order of the MA model, $c$ is the constant term of the ARMA model, $\phi_i$ is the free regression coefficient at time $t-i$, $\varepsilon_t$ is the error term at time $t$, and $\theta_j$ is the moving average coefficient at time $t-j$. The extended autocorrelation function (EACF) is selected to determine the order $p$ and $q$ of the AR model and the MA model through the maximum likelihood estimation method. The model parameters are estimated to build the ARMA model and predict the residual items.

Finally, the GCM predictions are obtained by adding the predictions of the linear model, MSSA and ARMA models.

### 4.2 Results and Analysis

This paper selects the time series of GCM from 1993 to Feb. 2017 provided by CSR for prediction research. Taking the data from January 1993 to the starting time of forecast as the training data, seven 2-year forecast experiments were carried out. The starting time was January, May and September 2013, January, May and September 2014 and January 2015, respectively. Through the analysis in Sect. 3, the MSSA window and main component were selected as 108 and 10. Through EACF analysis, it is found that ARMA(0,6), ARMA(0,11) and ARMA(3,9) have good applicability to the residual items in the X, Y and Z directions of GCM, respectively.

**Table 4.** The statistical precision of LM+MSSA+ARMA model (mm)

| Lead prediction | 6 month | | | 12 month | | | 24 month | | |
|---|---|---|---|---|---|---|---|---|---|
| | X | Y | Z | X | Y | Z | X | Y | Z |
| Max | 2.92 | 1.53 | 7.11 | 3.28 | 2.34 | 7.11 | 3.28 | 2.34 | 7.34 |
| Min | −2.34 | −2.19 | −6.23 | −2.34 | −2.19 | −6.60 | −4.80 | −2.19 | −6.60 |
| Mean | 0.14 | −0.15 | 0.58 | 0.08 | −0.14 | 0.86 | −0.32 | −0.07 | 0.91 |
| RMS | 1.29 | 1.03 | 3.29 | 1.35 | 1.08 | 3.45 | 1.53 | 1.08 | 3.46 |

Figure 9 shows the comparison between the predictions and the true values of GCM in the three directions. The black line in the figure represents the true values of GCM, and the red line represents the predicted values. The statistical results of the prediction precision of the LM+MSSA+ARMA model are shown in the Table 4. As shown in Fig. 8 and Table 4 that the combined model has the best prediction precision when the prediction length is 6 months. The RMSE in X, Y, and Z directions are 1.29 mm, 1.03 mm and 3.29 mm, respectively, and the precision of X and Y directions are all within 1.5 mm. The precision of the Z direction is relatively poor, which may be caused by the large change in the amplitude of GCM in the Z direction. As the prediction time increases, the predicted precision shows a downward trend as a whole, reaching the maximum when the prediction length is 2 years, and the prediction precision in the three directions is 1.53 mm, 1.08 mm and 3.46 mm respectively. In general, the LM+MSSA+ARMA model has good performance in short-term, medium-term and long-term predictions of GCM, and can provide prediction precision of 1.5 mm, 1.1 mm and 3.5 mm in the X, Y and Z directions respectively.

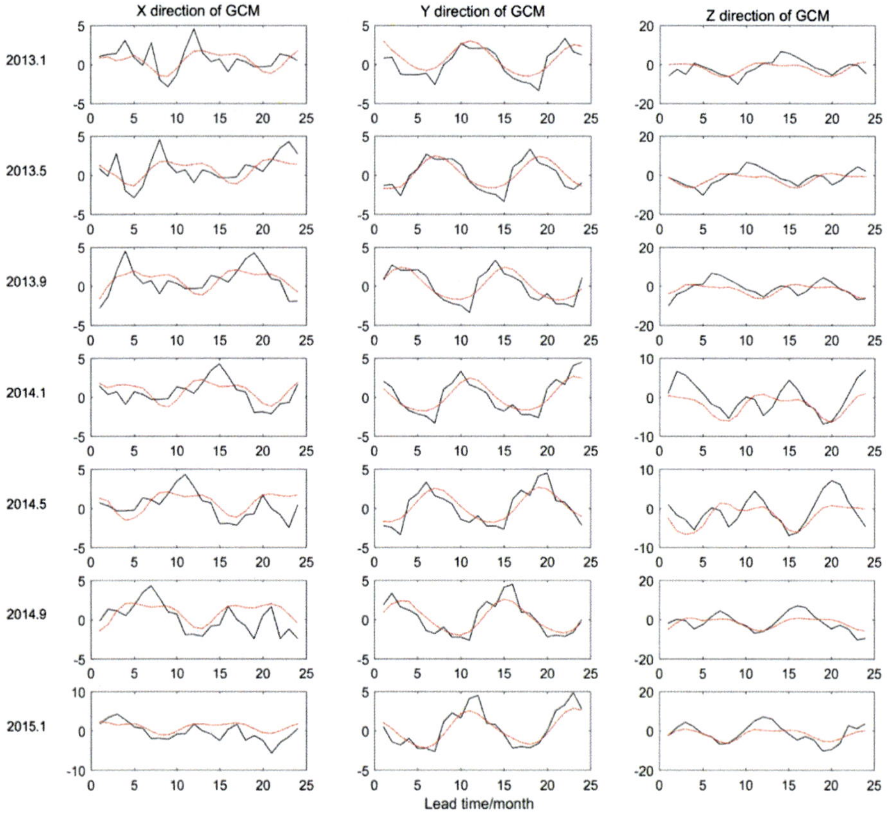


**Fig. 9.** Comparison of GCM predictions

# 5 Conclusions

The GCM products estimated from SLR data released by CSR are analyzed to determine the variation of geocenter motion. MSSA is used to analyze the GCM series in the X, Y and Z directions from 1993 to 2017.2. The seasonal variation periods are 0.99 years, 0.5 years, 0.58 years and 1.47 years, and the long period term is 6.09 years in the X, Y and Z directions, which shows that the MSSA can well display and amplify the main periodic signals of the original time series of GCM. The annual characteristics in the three directions are the most obvious and the wave oscillation is stable. The GCM in the X, Y and Z directions are directly analyzed by using MSSA to determine its trends. The long-term trends of the three directions are 0.05 mm/yr, 0.04 mm/yr and −0.10 mm/yr, respectively. The migration velocity in Z direction is obviously higher than that in X and Y directions, which may be mainly caused by the water mass redistribution in the northern and southern hemispheres.

To obtain high-precision geocentric motion prediction parameters, the method combining LM and MSSA and ARMA models is used to predict GCM with ahead 2 years. The results show that the LM+MSSA+ARMA model can effectively predict GCM parameters, and can provide prediction precision of 1.5 mm, 1.1 mm and 3.5 mm in X, Y and Z directions, respectively.

**Acknowledgements.** We are very grateful to CSR for providing GCM products (http://ftp.csr.utexas.edu/pub/slr/geocenter/). This research is supported by the National Natural Science Foundation of China (Grant Nos. 41774001 & 41704015), the Basic Science and Technology Project of China (Grant No. 2015FY310200).

## References

Altamimi, Z., Collilieux, X., Metivier, L.: ITRF2008: an improved solution of the international terrestrial reference frame. J. Geodesy **85**(8), 457–473 (2011). https://doi.org/10.1007/s00190-011-0444-4

Baur, O., Kuhn, M., Featherstone, W.E.: Continental mass change from GRACE over 2002–2011 and its impact on sea level. J. Geodesy **87**(2), 117–125 (2013). https://doi.org/10.1007/s00190-012-0583-2

Blewitt, G.: Self-consistency in reference frames, geocenter definition, and surface loading of the solid Earth. J. Geophys. Res. Solid Earth **108**(B2), 2103 (2003). https://doi.org/10.1029/2002JB002082

Blewitt, G., Clarke, P.: Inversion of Earth's changing shape to weigh sea level in static equilibrium with surface mass redistribution. J. Geophys. Res. Solid Earth **108**(B6), 2311 (2003). https://doi.org/10.1029/2002JB002290

Blewitt, G., Lavallee, D.E., Clarke, P.J., Nurutdinov, K.: A new global mode of Earth deformation: seasonal cycle detected. Science **294**(5550), 2342–2345 (2001). https://doi.org/10.1126/science.1065328

Bouille, F., Cazenave, A., Lemoine, J.M., Cretaux, J.F.: Geocentre motion from the DORIS space system and laser data to the Lageos satellites: comparison with surface loading data. Geophys. J. Int. **143**(1), 71–82 (2000). https://doi.org/10.1046/j.1365-246x.2000.00196.x

Cheng, M.K., Tapley, B.D., Ries, J.C.: Deceleration in the Earth's oblateness. J. Geophys. Res. Solid Earth **118**(2), 740–747 (2013). https://doi.org/10.1002/jgrb.50058

Cheng, M.K., Ries, J.C., Tapley, B.D.: Geocenter variations from analysis of SLR data. In: IAG Commission Symposium 2010, Reference Frames for Applications in Geosciences (REFAG2010), Mame-La-Vallee, France, 4–8 October 2010, pp. 19–25 (2013b). https://doi.org/10.1007/978-3-642-32998-2_4.

Crétaux, J., Soudarin, L., Davidson, F.J.M., Gennero, M., Bergé-Nguyen, M., Cazenave, A.: Seasonal and interannual geocenter motion from SLR and DORIS measurements: comparison with surface loading data. J. Geophys. Res. Solid Earth **107**(B12), 2374 (2002). https://doi.org/10.1029/2002JB001820

Dong, D., Dickey, J.O., Chao, Y., Cheng, M.K.: Geocenter variations caused by atmosphere, ocean and surface ground water. Geophys. Res. Lett. **24**(15), 1867–1870 (1997). https://doi.org/10.1029/97GL01849

Dong, D., Yunck, T., Heflin, M.: Origin of the international terrestrial reference frame. J. Geophys. Res. Solid Earth **108**(B4), 2200 (2003). https://doi.org/10.1029/2002JB002035

Golyandina, N., Nekrutkin, V., Zhigljavsky, A.: Analysis of Time Series Structure. Chapman and Hall/CRC, New York (2001). https://doi.org/10.1201/9780367801687

Feissel-Vernier, M., Le Bail, K., Berio, P., Coulot, D., Ramillien, G., Valette, J.J.: Geocentre motion measured with DORIS and SLR, and predicted by geophysical models. J. Geodesy **80**(8), 637–648 (2006). https://doi.org/10.1007/s00190-006-0079-z

Greff-Lefftz, M., Legros, H.: Fluid core dynamics and degree-one deformations: Slichter mode and geocenter motions. Phys. Earth Planet. Inter. **161**(3–4), 150–160 (2007). https://doi.org/10.1016/j.pepi.2006.12.003

Guo, J.Y., Han, Y.B., Hwang, C.W.: Analysis on motion of Earth's center of mass observed with CHAMP mission. Sci. China, Ser. G **51**(10), 1597–1606 (2008). https://doi.org/10.1007/s11433-008-0152-0

Guo, J.Y., Chang, X.T., Han, Y.B., Sun, J.L.: Periodic geocenter motion measured with SLR in 1993–2006. Acta Geodaetica et Cartographica Sinica **38**(4), 311–317 (2009). https://doi.org/10.3321/j.issn:1001-1595.2009.04.005

Hassani, H.: Singular spectrum analysis: methodology and comparison. J. Data Sci. **5**(2), 239–257 (2007). https://mpra.ub.uni-muenchen.de/4991/

Klemann, V., Martinec, Z.: Contribution of glacial-isostatic adjustment to the geocenter motion. Tectonophysics **511**(34), 99–108 (2011). https://doi.org/10.1016/j.tecto.2009.08.031

Kong, Q., Guo, J., Sun, Y., Zhao, C., Chen, C.: Centimeter-level precise orbit determination for the HY-2A satellite using DORIS and SLR tracking data. Acta Geophys. **65**(1), 1–12 (2017). https://doi.org/10.1007/s11600-016-0001-x

Kuzin, S.P., Tatevian, S.K.: Determination of seasonal geocenter variations from DORIS, GPS and SLR data. In: Proceedings of the Joumees Systemes de Reference Sptio-Temporels2005, Warsaw, pp. 76–77 (2005)

Kuzin, S.P., Tatevian, S.K., Valeev, S.G., Fashutdinova, V.A.: Studies of the geocenter motion using 16-years DORIS data. Adv. Space Res. **46**(10), 1292–1298 (2010). https://doi.org/10.1016/j.asr.2010.06.038

Metivier, L., Greff-Lefftz, M., Altamimi, Z.: On secular geocenter motion: the impact of climate changes. Earth Planet. Sci. Lett. **296**(3–4), 360–366 (2010). https://doi.org/10.1016/j.epsl.2010.05.021

Montag, H.: Geocenter variations derived by different satellite methods. IERS Technical Note25. Paris: Observatoire de Paris, 71–74 (1999)

Oropeza, V., Sacchi, M.: Simultaneous seismic data denoising and reconstruction via multichannel singular spectrum analysis. Geophysics **76**(3), V25–V32 (2011). https://doi.org/10.1190/1.3552706

Pavlis, H.: Fortnightly resolution geocenter series: a combined analysis of Lageos 1 and 2 SLR data (1993–96). IERS Technical Note 25, Paris, Observatoire de Paris, pp. 75–84 (1999)

Pearlman, M., et al.: Laser geodetic satellites: a high-accuracy scientific tool. J. Geodesy **93**(11), 2181–2194 (2019). https://doi.org/10.1007/s00190-019-01228-y

Riddell, A.R., King, M.A., Watson, C.S., Sun, Y., Riva, R.E.M., Rietbroek, R.: Uncertainty in geocenter estimates in the context of ITRF2014. J. Geophys. Res. Solid Earth **122**, 4020–4032 (2017). https://doi.org/10.1002/2016JB013698

Ries, J.C.: Reconciling estimates of annual geocenter motion from space geodesy. In: Proceedings of the 20th International Workshop on Laser Ranging, Potsdam, Germany, pp. 10–14 (2016)

Rietbroek, R., Brunnabend, S.E., Kusche, J., Schroter, J.: Resolving sea level contributions by identifying fingerprints in time-variable gravity and altimetry. J. Geodyn. **s59-60**(5), 72–81 (2012a). https://doi.org/10.1016/j.jog.2011.06.007.

Rietbroek, R., et al.: Global surface mass from a new combination of GRACE, modelled OBP and reprocessed GPS data. J. Geodyn. **59–60**(5), 64–71 (2012b). https://doi.org/10.1016/j.jog.2011.02.003

Shen, Y., Guo, J., Liu, X., Wei, X., Li, W.: One hybrid model combining singular spectrum analysis and LS+ARMA for polar motion prediction. Adv. Space Res. **59**(2), 513–523 (2017). https://doi.org/10.1016/j.asr.2016.10.023

Shen, Y., Guo, J., Liu, X., Kong, Q., Guo, L., Li, W.: Long-term prediction of polar motion using a combined SSA and ARMA model. J. Geodesy **92**(3), 333–343 (2018). https://doi.org/10.1007/s00190-017-1065-3

Sun, Y., Riva, R., Ditmar, P.: Optimizing estimates of annual variations and trends in geocenter motion and J(2) from a combination of GRACE data and geophysical models. J. Geophys. Res. Solid Earth **121**(11), 8352–8370 (2016). https://doi.org/10.1002/2016JB013073

Trupin, A.S., Meier, M.F., Wahr, J.M.: Effect of melting glaciers on the Earth's rotation and gravitational field. Geophys. J. Int. **108**(1), 1–15 (1992). https://doi.org/10.1111/j.1365-246X.1992.tb00835.x

Vautard, R., Ghil, M.: Singular spectrum analysis in nonlinear dynamics, with applications to paleo-climatic time series. Physica D-Nonlinear Phenomena **35**(3), 395–424 (1989). https://doi.org/10.1016/0167-2789(89)90077-8

Wang, X., Cheng, Y., Wu, S., Zhang, K.: An enhanced singular spectrum analysis method for constructing nonsecular model of GPS site movement. Journal of Geophysical Research: Solid Earth **121**(3), 2193–2211 (2016). https://doi.org/10.1002/2015JB012573

Watkins, M.M., Eanes, R.J.: Observations of tidally coherent diurnal and semidiurnal variations in the geocenter. Geophys. Res. Lett. **24**(17), 2231–2234 (1997). https://doi.org/10.1029/97GL52059

Wei, N., Shi, C., Liu, J.N.: Effects of surface loading and heterogeneous GPS network on Helmert transformation. Chin. J. Geophys. **59**(2), 484–493 (2016). https://doi.org/10.6038/cig20160208

Wu, X., et al.: Simultaneous estimation of global present-day water transport and glacial isostatic adjustment. Nat. Geosci. **3**(9), 642–646 (2010). https://doi.org/10.1038/ngeo938

Wu, X., Ray, J., Dam, T.V.: Geocenter motion and its geodetic and geophysical implications. J. Geodyn. **58**(3), 44–61 (2012). https://doi.org/10.1016/j.jog.2012.01.007

Zhao, C., Ma, T.: The estimation and prediction of geocenter motion based on GNSS/SLR weekly solutions. In: 42nd COSPAR Scientific Assembly, Pasadena, California, USA, 14–22 July 2018, B2-1 (2018)

Zhao, C., Qiao, L., MA, T.: Estimation and prediction of geocenter motion based on GNSS weekly solutions of IGS. In: American Geophysical Union, Fall Meeting, San Francisco, California, 9–13 December 2019, G12A-05 (2019)

Zhou, M.S., Guo, J.Y., Shen, Y., Kong, Q.L., Yuan, J.J.: Extraction of common mode errors of GNSS coordinate time series based on multi-channel singular spectrum analysis. Chin. J. Geophys. **61**(11), 4383–4395 (2018). https://doi.org/10.6038/cjg2018L0710

Zotov, L., Bizouard, C.H., Shum, C.K.: About possible interrelation between Earth rotation and climate variability on a decadal time-scale. Geodesy Geodyn. **7**(3), 216–222 (2016). https://doi.org/10.1016/j.geog.2016.05.005

# Monitoring of the Turag River Dyke in Dhaka City Using Ground Penetrating Radar (GPR): A New Approach of Flood Risk Reduction in Bangladesh

Reshad Md. Ekram Ali[1](✉), Fannian Kong[2], Rajinder Kumar Bhasin[2], Salma Akter[1], Mohammad Zohir Uddin[1], and Shahtaj Karim[1]

[1] Geological Survey of Bangladesh, 153 Pioneer Road, Segunbagicha, Dhaka 1000, Bangladesh
[2] Norwegian Geotechnical Institute, Ullevaal Stadion, P.O. Box 3930, 0806 Oslo, Norway

**Abstract.** The focus of this study is placed on inspecting the Turag River dyke in Dhaka by using the Ground Penetrating Radar (GPR) which is often damaged during the monsoon season due to erosion caused by flooding and overtopping. The dyke was inspected to identify the weak zones at several locations by using GPR which is a non-destructive investigation technique. It requires no physical penetration into the ground, can perform quick survey and hence low cost involved compared to other surveys.

GPR survey has been conducted in three selected stretches of the dyke. These selected stretches are Mirpur Sluice Gate, Rostampur Cross Dyke and Prottasha Bridge areas. GPR investigations on these particular portions of the dyke have done to monitor the health of the dyke and to locate any disturbed or weak zones which may need strengthening before monsoon.

Radargram of Mirpur Sluice Gate area shows four distinct reflection layers of pitch (tar) mixed with brick; brick chips with sand; pre-existing footpath and compacted soil. Compacted soil below the pre-existing footpath formed by silty clay to clay materials found with slight variation of materials types, materials compactness and wetness, indication comparative weak zones in silty clay to clay. The lines representing the reflections from the ground layers suddenly end from both sides in the radargram indicating the existence of the sluice gate. In Rostampur Cross Dyke three subsurface layers can easily be identified from the radargram. The uppermost unit is mainly composed of brick chips and silty clay materials. Below the top layer, lithology of second layer has not been identified. Underneath the second layer the main dam materials are silty clay to clay. Variation of materials, compactness and wetness are also observed here. Radargram from GPR measurement on the highway beside Prottasha Bridge clearly shows that the bridge structure of the section is different from the adjacent sections at both sides. Reflections from pipes or cables under the highway are strongly visible. Although dyke materials are comparatively fine textured and wet but results were satisfactory and encouraging.

**Keywords:** GPR · Dyke · Radargram · Compactness


H. Khabbaz et al. (Eds.): GeoChina 2021, SUCI, pp. 46–57, 2021.
https://doi.org/10.1007/978-3-030-79650-1_4

## 1 Introduction

Dhaka is the capital and the largest city in Bangladesh. It is situated in the South Asia and the world's 11th largest megacity (UN 2016) with a population of 18.2 million living in an area of 1528 km$^2$ (Swapan et al. 2017) Presently, the city faces a number of environmental and other hazards such as river flooding, water logging, earthquakes, fire hazards, traffic congestion, etc. The city is surrounded by the Buriganga River to the south, the Turag River to the west, the Tongi khal to the north, and the Balu River to the east (Fig. 1). The city and adjoining areas are composed of alluvial terraces of the southern part of the Madhupur tract and low-lying areas of floodplain deposits. The elevation of Greater Dhaka ranges between 2 to 13 m AMSL and most of the urbanized areas are between 6 to 8 m covers about 75 km$^2$, while 170 km$^2$ of Greater Dhaka is below 6 m above mean sea level (JICA 1987). Flooding is the best-known hazard of Dhaka. In 1988 most of Dhaka has been inundated by prolonged record-setting flood, like other floods, which were clearly visible on space satellite photographs (Rasid 1993a).

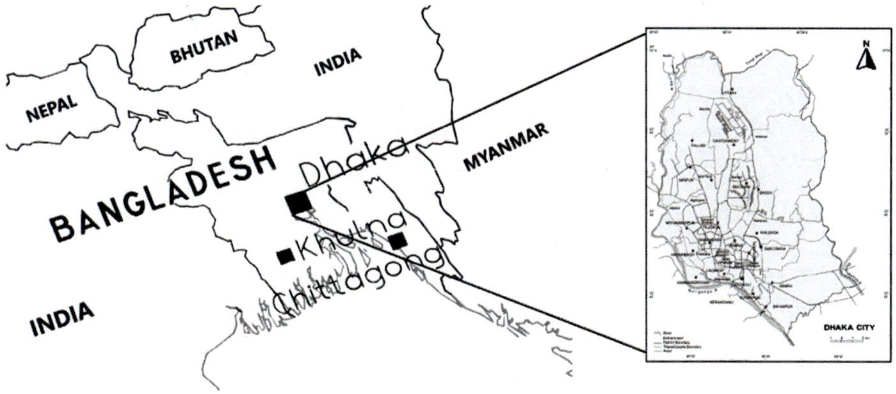


**Fig. 1.** Location map of Dhaka City

The main causes of flooding in Dhaka City are spill over water from the surrounding rivers and internal water logging. In recent history, Dhaka city experienced major floods in 1954, 1955, 1970, 1974, 1980, 1987, 1988, and 1998 due to the overflow of surrounding rivers. Among these, the 1988 and 1998 floods were catastrophic (Ahmed 1999; Chowdhury et al. 1998; Islam and Chowdhury 2002; Islam et al. 2008). Prof. Hye (1999) carried out a rapid appraisal of flood-affected people during the flood, dividing the flood-affected area into three categories: most severely affected area, severely affected area, and moderately affected area. It was found that the flood displaced or dislocated 94% of families in the most severely affected areas, 52% of families in severely affected areas, and 50% in moderately affected areas. It was also estimated that the total flood-affected population would likely be 4.55 million.

Ground Penetrating Radar (GPR) is a useful tool for geophysical survey, it has been using to solve different geological problems for long time (Doolittle and Collins 1995; Doolittle et al. 2006; Shih and Doolittle 1984; Benson 1995). Similarly, GPR can be used to detect weak zones in the dyke that could be breached easily in the near future. GPR has similar record as seismic survey (Jol 2009; Baker et al. 2007). The main difference between GPR and seismic methods is that seismic reflection comes from a boundary where there is an acoustic impedance contrast and in the case of GPR the reflection comes from a boundary where there is a dielectric constant contrast. However, in seismic reflection method, the depth of penetration is very large so the geoscientists think about gross lithology but in case of GPR they have to think very minutely about the lithology and other variation within a very shallow depth in the range of 10 m. For any geophysical methods, the detection resolution is a portion (for example, one tenth) of the test distance. Therefore, GPR normally gives detection results with much higher resolution than the seismic method. The other advantages of using GPR are that for detection using electromagnetic wave, the sensors (or antennas) does not need to have good contact with the ground, which enables a fast survey speed, and that the GPR signal is generated by electronics, and is coherent and repetitive, which allows to use advanced signal and image processing methods. The most significant performance limitation of GPR is in high-conductivity materials such as clay soils and soils that are salt contaminated. Performance is also limited by signal scattering in rocky soils. Compare to the resistivity survey GPR can penetrates subsurface very little where the subsurface is conductive. Whereas, the main advantage of GPR over resistivity survey is that it can be run over hard surface that is not penetrable by electrode and resolution of radargram is very high compared to other geophysical methods.

## 2 Floods Mitigation in Dhaka

In 1864, the Buckland Flood Protection Embankment along the Buriganga River was the first attempt to mitigate flood damage in Dhaka City. Plans for comprehensive flood protection for Greater Dhaka have been under study and consideration for many years, but the catastrophic flooding that occurred in 1987 and 1988 brought into focus the urgent need to proceed with immediate action. In 1989, construction activities commenced on a "crash program" embank the western part of the city along the Turag River. It provides flood protection facilities to the western half of Dhaka and includes the most highly urbanized areas, covering about 87% of the population. This mitigation project has also encouraged urban invasion of the floodplain in an area (Rasid and Azim Mallik 1993b; Rasid and Mallik 1993; Rasid and Azim Mallik 1996; JICA 1991a). Presently, inside the embankment rapid urban expansion is continuing by both private housing companies and Rajdhani Unnayan Kartripakkha (RAJUK), important RAJUK projects are Uttara Residential Model Town Projects, especially 3rd phase, Uttara Apartment Project, etc. (http://www.rajukdhaka.gov.bd/rajuk/webHome). Although, the construction and maintenance costs and the operational requirements of the entire scheme are very heavy for the country. Unfortunately, the eastern part of Dhaka City is still unprotected.

Dhaka West have been encircled by embankments, floodwalls, raised roads to give protection against riverine flood (JICA 1991b). Approximately there are 30 km of earthen embankment along Tongi khal, Turag River and Buriganga River. A total of 11 regulators are at the outfall of khals to the surrounding rivers along the embankment. One regulator and 12 sluice gates on the khals at the crossings with Biswa Road, DIT Road, Pragati

Sarani, Mymensingh Road, and railway line at uttar khan. One pumping station is at the outfall of Kallyanpur khal to the Turag River and another one at the outfall of Dholaikhal to the Buriganga River. These pump stations are for draining rainwater from parts of Dhaka West. A special 10.53 km embankment is to surround the Zia International Airport.

After Implementation of the flood-control project in Dhaka West, unplanned and uncontrolled expansion of urban areas stretched rapidly toward the low-lying areas and floodplains adjacent to the flood-protection embankment and river. Evaluation of Phase I work revealed that the existing earthen embankment is unstable and cracking in large sections, though some parts have recently been stabilized and other parts of the earthen embankment have been converted to roads. Construction of embankments through low-lying areas without providing adequate drainage facilities has caused internal flooding, adversely affecting the residents in those areas (Huq and Alam 2003). Residents of these lowlands suffer from inundation due to accumulation of rainwater after heavy rainfall. Land development through land-filling processes in the low-lying areas is causing a drastic reduction in water storage areas. Floods are one of the main natural hazards affecting the city and are associated with river water overflow and rain water stagnation.

Performance of western dykes reveals that the design height of dykes was not adequate to tolerate the flood of 1998. Moreover, coordinated efforts from the organizations like Dhaka Water Supply and Sewerage Authority (DWASA), Bangladesh Water Development Board (BWDB) and Dhaka City Corporation (DCC) were not observed. As a result, some culverts and regulators along the raised roads were left open in the beginning of the flood of 1998. In 2004 condition flood protection dykes have deteriorated due to settlement activities around it. Internal flooding is surfacing gout as one of the disasters for protected Dhaka West in 2004 (Bala et al. 2009).

Moreover, the dykes were not properly maintained since its construction. Breaches of dykes may cause tremendous dame of properties and immense suffering of people. Now monitoring and maintenance of dyke are the prime important to protect Dhaka west from future floods.

## 3 Principle of GPR and River Dyke Monitoring

There are two types of the material properties, which have effects on the electromagnetic wave propagating inside the material. One is the relative dielectric constant $\varepsilon_r$, and the other is the resistivity. Tables 1 and 2 show respectively the dielectric constants $\varepsilon_r$ and the resistivities R for common ground materials.

Wave speed and propagation attenuation are related to the ground material properties. The wave velocity can be written as

$$v = \frac{2\pi f}{k_{real}} = \frac{1}{\sqrt{\mu\varepsilon_0}\sqrt{\varepsilon_r}} = \frac{c}{\sqrt{\varepsilon_r}} \tag{1}$$

Where $c$ is the speed of light in air and $c = 0.3$ m/ns. The list of wave speeds in ground materials shown in Table 1 is calculated by using the above equation.

A general rule is that for detecting deep targets one uses low frequency and long antennas and detection distance is 10 times of the antenna length. The above empirical relationship can help to choose antenna lengths according to the detection depth.

**Table 1.** Dielectric constants measured at 100 MHz of common earth materials (Cardimona 2002).

| Material | Typical dielectric constant | Radar propagation velocity (m/ns) |
|---|---|---|
| Air | 1 | 0.30 |
| Water | 81 | 0.033 |
| Granite | 9 | 0.10 |
| Limestone | 6 | 0.12 |
| Sandstone | 4 | 0.15 |
| Rocks | 4-12 | 0.15-0.087 |
| Dry sand | 4-6 | 0.15-0.12 |
| Wet sand | 30 | 0.055 |
| Dry clay | 8 | 0.11 |
| Wet clay | 33 | 0.052 |
| Dry soils | 3-8 | 0.17-0.11 |
| Wet soils | 4-40 | 0.15-0.047 |
| Asphalt | 3-6 | 0.17-0.12 |
| Concrete | 9-12 | 0.10-0.087 |

**Table 2.** Typical ranges of electrical resistivity and conductivity of earth material (Samouëlian et al. 2005)

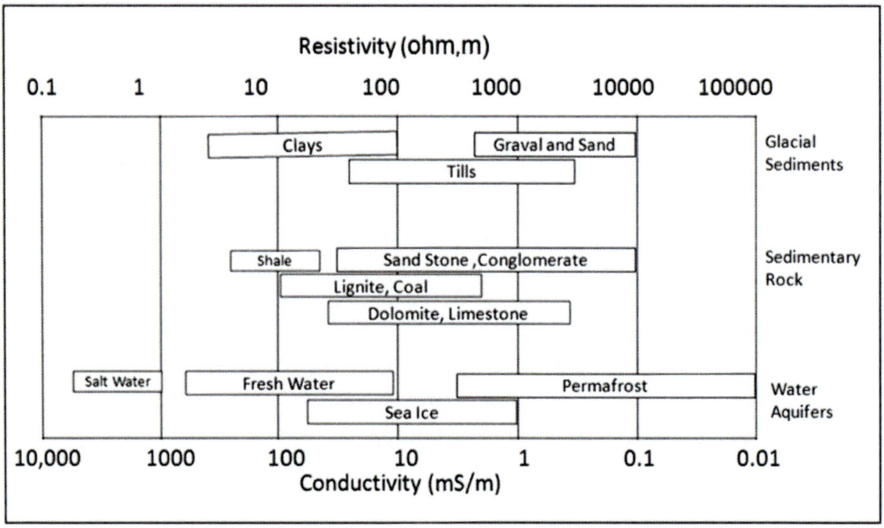

The resolution is also related to the wavelength used. Here we adapt the criterion that the detection resolution is ¼ wavelengths, i.e., half of the antenna length. Hence our 10 cm antenna is best suited for detection distance 1 m with a resolution 5 cm, while our 1.2 m antenna is best suited for detection distance 12 m with a resolution 60 cm.

The conventional GPR test mode uses the reflection wave, where two antennas with a fixed spacing are used, one for transmitting and the other for receiving signals. They are moved along the river dyke. Electromagnetic waves can be transmitted through the ground with the help of the transmitting antenna. The energy of the wave is reflected back towards the surface where the electric or magnetic properties change (due to the weak zones etc.) in the river dyke. A second antenna receives the reflected wave. The ground materials will attenuate the radar wave. The detection distance is therefore dependent on the types of river dyke materials. GPR reflection measurement on top of the river dyke is a commonly accepted method used for the dyke monitoring (Fig. 2).

**Fig. 2.** Layout of GPR (1.2 m) test where a pair of 1.2 m antennas are used.

## 4 Results and Discussion

The dyke is located in the north and north-western part of the Dhaka city on Turag dyke. This dyke is approximately 30 km of earthen embankment along Tongi khal, Turag River, and Buriganga River. Three locations (Fig. 3) on the Turag dyke have been selected for the study based on radargrams with the objectives to detect weak zones due to fractures/cracks, organic materials and void spaces.

**Mirpur Sluice Gate Area**

Mirpur Sluice Gate area have been selected for GPR survey on Tongi Dyke (Fig. 3). The starting and ending positions of data acquisition are 23°49′48″N, 90°49′44″E to 23°49′44″N, 90°20′39″E. The main purpose of this survey is to locate the sluice gate zone with GPR. Figure 4 shows a radargram across the measurement line.

For collecting a radargram in the sluice gate area a pair of 1.2 m length antennas such as shown in Fig. 4 have been used with 1 m spacing between receiving and transmitting antennas.

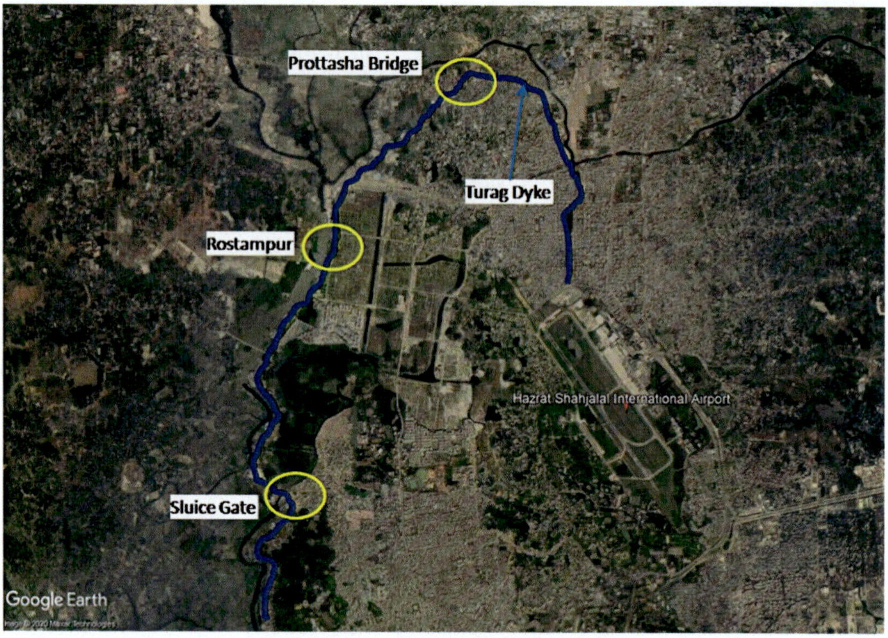


**Fig. 3.** Location map of 3 test study areas (Sluice Gate, Rostampur and Prottasha Bridge areas) on the Turag dykes in Dhaka.

In Fig. 4 there are four distinct reflections observed in the radargram: the 1st is the reflection at the interface between air and layer 1, the 2nd reflection is at the interface between layer 1 and layer 2, the third reflection is at the interface between layer 2 and layer 3 and the fourth reflection is at the interface between layer 3 and layer 4.

It is certain from our survey that the uppermost layer (layer 1) represents combine pitch (tar), brick, brick chips with sand. The layer 2 may represent the pre-existing footpath over which new pitch road has been made. The third layer may represent compacted soil below the pre-existing footpath. Below the third layer earth fill that is the layer 4 formed by silty clay to clay materials found with slight variation of materials, compactness and wetness. Results of GPR survey, confirmed by digging and hand auger where necessary.

Each division of the vertical axis is 20 ns representing a depth of 1 m with the assumption of the wave speed 10 cm/ns. So the approximate depth of layers 1, 2 and 3 are about 0.5 m each. Due to their thinness, it is very hard to differentiate among pitch, brick and brick chips zone within the uppermost layer (layer 1). Horn antenna or 60 cm length antenna may give good result for identification of pitch layer in the uppermost layer.

In the radargram shown in Fig. 5, two sets of anomaly are found. One is between the trace numbers 355 to 380 and the other is in between 320 to 340. The anomaly in the red circle represents the sluice gate. The lines representing the reflections from the ground layers suddenly end from both side of red circle indicating the existence of the sluice gate. The anomaly in between trace numbers 320 to 350 shows a different reflection

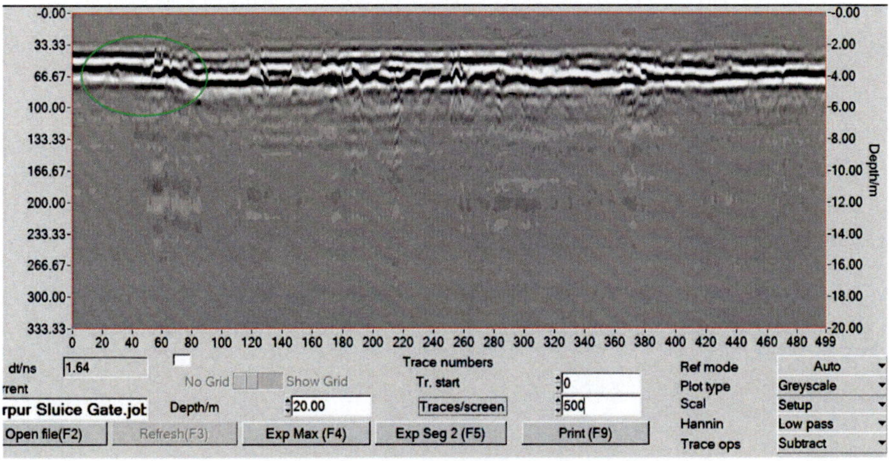


**Fig. 4.** Radargram Over Mirpur Sluice Gate.

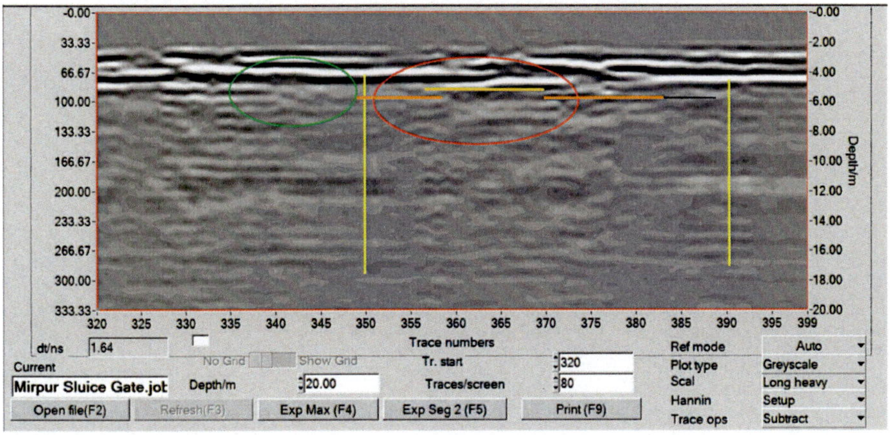


**Fig. 5.** The Sluice gate zone in between yellow mark (Red Circle).

pattern from the other one. This discontinuity of signal may be due to the beginning part of the sluice gate structure.

**Rostampur Area**

A GPR Survey was performed in Rostampur Cross dyke (23°52′32″ N, 90°21′01″ E) to identify subsurface lithology (Fig. 3) and the test result is shown in Fig. 6. In this survey the 1.2 m antenna has been used. Between the trace nos. 1 and 110, three lithological layers can easily be identified from the radargram. The uppermost unit is mainly composed of brick chips and silty clay materials with a thickness 0.5 m. Below the top layer there is another layer of thickness 0.5 m. The lithology of this layer has not been identified. Below the second layer the main dam materials silty clay to clay are found. From trace no. 111 to the end of the radargram, we can only see two layers.

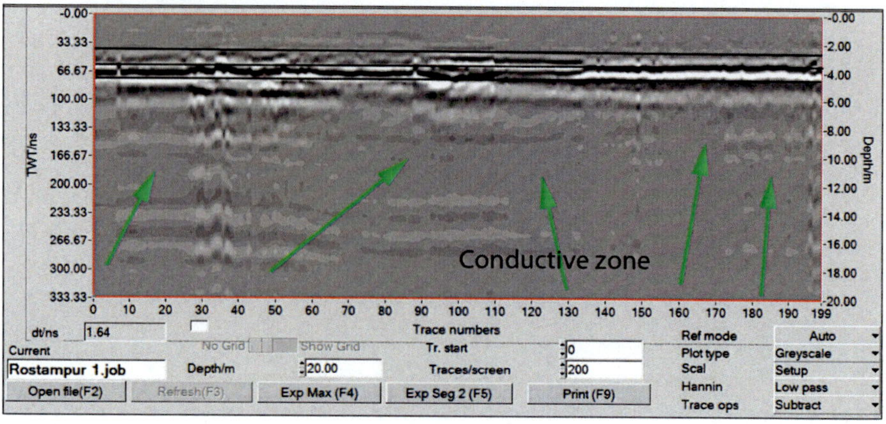


**Fig. 6.** Test result in Rostampur area.

## Prottasha Bridge and on Highway Nearby

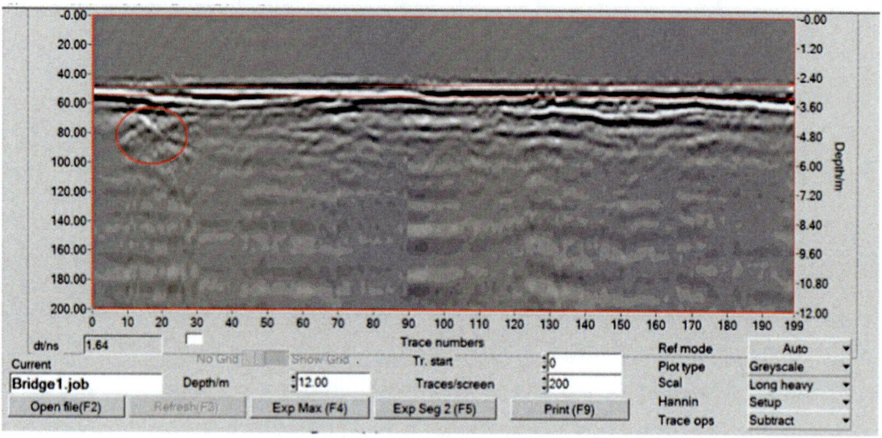


**Fig. 7.** Radargram along connecting road of Prottasha bridge.

Figure 7 shows the GPR survey at Prottasha Bridge (23°54′21″ N, 90°22′35″ E) (Fig. 3) using the 1.2 m antenna used in all of the outdoor tests. In Fig. 7, the uppermost red line represents the response from base of pitch and stone chips. The 2$^{nd}$ red line marks the response from base of brick and below this the whole material might be silty clay.

The area within trace nos. 10–25 marked by the red circle may indicate the junction between the highway and the bridge connecting road which can be identified by the criss-crossed nature of the signal. Here, on the sloping side of the highway, new materials have been dumped for the bridge connecting road construction. From trace nos. 133–173 the signal shows a concave nature. The area within this section might be more compacted compared to its adjacent sides.

Figure 8 shows the radargram for GPR measurement on the highway beside Prottasha bridge. From this figure one can clearly see that the ground structure of the section between two yellow lines (trace nos. 125 to 265) is different to the adjacent sections at both sides. This is the section when the GPR moves close to the bridge connecting road. The red lines show the reflections from a shallow layer (brick to sand interface?). At trace nos. 30, 85, 330, there may be reflections from pipes or cables under the highway.

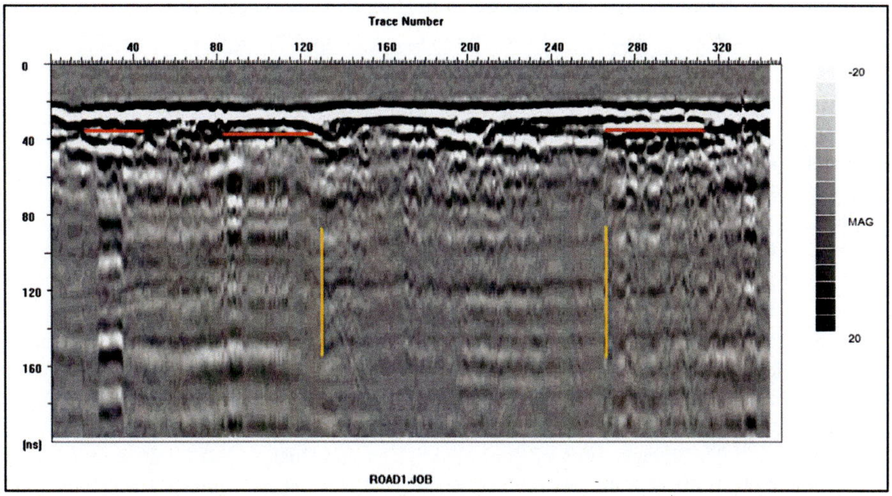


**Fig. 8.** Results of GPR measurement on the highway beside Prottasha bridge

## 5 Conclusions

GPR provides an efficient and resourceful means for shallow subsurface studies. Radargram of GPR show that on top of the dyke there is 6 cm thick asphalt carpeting with stone chips. Below the asphalt layer there is 13 cm thick compacted layer of brick chips mixed with sand (Locals inform maximum thickness of the brick chips layer is about 40 cm). Below brick chips there is sand. There is no brick soling below the brick chips layer. GPR test using 1.2 m antenna cannot resolve the layers with thickness 6 cm and 13 cm. However, Figs. 3 and 5 show that using 10 cm antennas can detect those layers.

GPR can be used to detect the material discontinuities along the measurement direction (e.g., see Figs. 3 and 7). Those discontinuity locations may suggest that the ground is built by different materials or the ground is deformed, and may indicate weak locations of a dyke.

In the Turag River dike the weak zones include loosely compacted layers, joints in different ground materials within the dykes, presence of highly compressive layers, etc. Generally speaking, materials such as concrete, sand, rock, gravel, sandy soil etc. are good materials for radar wave to penetrate. However, silty clay and wet clay etc. are the difficult materials for radar wave to penetrate. For good materials the GPR detection distance can be longer than 10 m. Even though the detection distance is limited in clay material etc.

The tests show that GPR can make quick and continuous measurements on river dyke etc. GPR can provide high-resolution image of the ground layering information. Moreover, GPR survey is a non-destructive investigation method, as well as being quick. Survey speed can be as fast as 2–3 km/hour. For an experienced operator, the interpretation of the survey results can be made on site without further data processing (inversion etc.). Hence, the GPR survey is low cost geotechnical investigation method.

The western part of Dhaka city is protected from river flooding by an encircling embankment. During most of the monsoon period, the water level of the river remains higher than the water level inside the city area. Adaptations to climate change and to economic development are important drivers in designing flood management schemes. Flood risk management is strongly related to spatial planning: the location of new developments, flood proof structures and space for rivers. Risk communication is a valuable approach to promote flood awareness and improve agile flood mitigation systems.

## References

Ahmed, F.: Impact on water supply sanitation and waste management. In: Nishat, A., Reazuddin, M., et al. (eds.) The 1998 Flood: Impact on Environment of Dhaka City. Department of Environment in conjunction with IUCN Bangladesh (1999)

Baker, G.S., Jol, H.M. (eds.): Stratigraphic analyses using GPR, Geological Society of America Special Papers, vol. 432, 181 p. The Geological Society of America, Boulder, CO (2007)

Bala, S.K., et al.: Performance of flood control works around Dhaka City during major floods in Bangladesh, Institute of Water and Flood Management, Bangladesh University of Engineering. In: 2nd International Conference on Water & Flood Management (2009)

Benson, A.K.: Applications of ground penetrating radar in assessing some geological hazards: examples of groundwater contamination, faults, cavities. J. Appl. Geophys. **33**(1–3), 177–193 (1995)

Cardimona, S.: Subsurface investigation using ground penetrating radar. In: 2nd International Conference on the Application of Geophysical and NDT Methodologies to Transportation Facilities and Infrastructure, Los Angeles, California (2002)

Chowdhury, J.U., Rahman, R., Bala, S.K., Islam, A.S.: Impact of 1998 Flood of Dhaka City and Performance of Flood Control Works Dhaka: Institute of Flood Control & Drainage Research, Bangladesh University of Engineering & Technology (1998)

Doolittle, J.A., Collins, M.E.: Use of soil information to determine application of ground penetrating radar. J. Appl. Geophys. **33**(1–3), 101–108 (1995)

Doolittle, J.A., Jenkinson, B., Hopkins, D., Ulmer, M., Tuttle, W.: Hydropedological investigations with ground penetrating radar (GPR): estimating water-table depths and local ground-water flow pattern in areas of coarse-textured soils. Geoderma **131**(3–4), 317–329 (2006). http://www.rajukdhaka.gov.bd/rajuk/webHome

Huq, S., Alam, M.: Building Safer Cities: The Future of Disaster Risk Flood Management and Vulnerability of Dhaka City (2003)

Hye, A.: Effect on livelihood. In: Nishat, A., Reazuddin, M., et al. (eds.) The 1998 Flood: Impact on Environment of Dhaka City. Department of Environment in conjunction with IUCN Bangladesh (1999)

Islam, A.S., Chowdhury, J.U. (eds.): Hydrological Characteristics of the 1998 flood in major rivers, Engineering Concern of Flood. Bangladesh University of Engineering and Technology, Dhaka (2002)

Islam, A.S., Haque, A., Bala, S.K.: Hydrological Aspects of Flood 2007. Dhaka, Bangladesh: Institute of Water and Flood Management (IWFM), Bangladesh University of Engineering and Technology (BUET) (2008)

JICA (Japan International Cooperation Agency): "Study on Storm Water Drainage System Improvement Project in Dhaka City." Supporting Report, Local Government Division of Ministry of Local Government Rural Development and Cooperative, Dhaka (1987)

JICA: Master Plan for Greater Dhaka Protection Project (Study in Dhaka Metropolitan Area), FAP 8A, Main Report and Supporting Reports I and II. Flood Plan Coordination Organization. Dhaka: Japan International Cooperation Agency (1991a)

JICA: Master Plan for Greater Dhaka Flood Protection Project. Flood Action Plan 8A, Main Report and Supporting Reports I and II. Flood Plan Coordination Organization (Presently WARPO), Dhaka (1991b)

Jol, H.M. (ed.): Ground Penetrating Radar: Theory and Applications, 524 p. Elsevier, Amsterdam (2009)

Rasid, H.: Areal extent of the 1988 flood in Bangladesh: how much did the satellite imagery show? Natural Hazards **8**(2), 189–200 (1993a)

Rasid, H.: Preventing flooding or regulating flood levels? Case studies on perception of flood alleviation in Bangladesh. Natural Hazards **8**(1), 39–57 (1993b)

Rasid, H., Azim Mallik, A.: Poldering vs compartmentalization: the choice of flood control techniques in Bangladesh. Environ. Manage. **17**(1), 59–71 (1993)

Rasid, H., Azim Mallik, A.: Living on the edge of stagnant water: an assessment of environmental impacts of construction-phase drainage congestion along Dhaka City Flood Control Embankment in Bangladesh. Environ. Manage. **12**(1), 89–98 (1996)

Samouëlian, A., Cousin, I., Tabbagh, A., Bruand, A., Richard, G.: Electrical resistivity survey in soil science: a review. Soil Tillage Res. **83**(2), 173–193 (2005)

Shih, S.F., Doolittle, J.A.: Using radar to investigate organic soil thickness in the Florida everglades. Soil Sci. Am. J. **48**(3), 651–656 (1984)

Swapan, M.S.H., Zaman, A.U., Ahsan, T., Ahmed, F.: Transforming urban dichotomies and challenges of South Asian megacities: rethinking sustainable growth of Dhaka, Bangladesh. Urban Sci. **1**, 31 (2017). https://doi.org/10.3390/urbansci1040031

UN: The World's Cities in 2016. World Urbanization Prospects: Data Booklet (ST/ESA/SER.A/392); United Nations, New York (2016)

# Relationships Between Compressive and Splitting Tensile Strengths of Cast and Core High-Strength Concrete Cylinders

Richard Campos[1(✉)], Matias M. Mendez Larrain[1], Musharraf Zaman[2], and Victor Pozadas[1]

[1] Roca Engineering, Inc., 3200 NW 38th Street, Oklahoma City, OK 73112, USA
richard.campos@ou.edu, {mmendez,vpozadas}@rocaengineering.com
[2] Department of Civil Engineering and Environmental Science, University of Oklahoma, 202 Boyd Street, Rm. 334, Norman, OK 73109, USA
zaman@ou.edu

**Abstract.** Distressed pavements are costly due to frequent maintenance, rehabilitation, and destructive testing. Furthermore, distressed pavements can lead to increased vehicle repair costs, safety concerns, and fatalities. The present study uses the splitting tensile and compressive strengths of 140 cast and core specimens to develop theoretical regression models that can better correlate the indirect splitting tensile and compressive strengths at 28 days for high-strength concrete. The laboratory experiments that were conducted in this study include splitting tensile and compressive strength tests using a universal testing machine. The reliability of the proposed models was verified by comparing their predictive capabilities with the laboratory tests. Statistical regression software was used to analyze the cast and core strength test results and develop regression models. The root mean square error (RMSE) was used to assess the accuracy between the values predicted by the regression models and laboratory values. Existing regression models reported by researchers and institutions were also analyzed with the RMSE method to determine the most accurate model. Based on the analysis, a positive nonlinear correlation between the splitting tensile strength and the compressive strength of high strength concrete was determined. The proposed regression equation exhibits small errors when compared to the laboratory results, which allow for efficient and accurate predictions of the splitting tensile strength.

**Keywords:** Splitting tensile strength relationship · Compressive strength relationship · Core and cast specimen relationship

## 1 Introduction

The assessment of a pavement's strength capacity is necessary as the pavement ages and sustains distresses. Pavement distresses can lead to costly vehicle repairs, vehicle crashes, and fatalities (U.S. Department of Transportation 2015; ASCE 2017; Justo-Silva


H. Khabbaz et al. (Eds.): GeoChina 2021, SUCI, pp. 58–69, 2021.
https://doi.org/10.1007/978-3-030-79650-1_5

and Ferreira 2019). The stress-inducing factors that degrade and crack pavements are temperature, vehicle traffic, and poor subgrade characteristics (Yoder and Witczak 1975). Due to the necessity of evaluating pavement performance by means of Quality Assurance and Quality Control (QA/QC) and ensuring driver safety, Non-Destructive Tests (NDT) and Semi-Destructive Tests (SDT) have been developed (Chen and Won 2007). The most common SDT method for determining the strength of an existing concrete pavement is by core testing, a method by which core specimens are extracted from the pavement and tested in laboratory. Similar to cast cylinder strength tests, core cylinders are also used to determine the compressive strength and splitting tensile strength (ASTM 2018c). Coring is a relatively easy process; however, there are more variables that can affect the strength of a core specimen, such as core size, moisture, drilling direction, and core damages (Chen et al. 2013). Consequentially, there exists no relationship between the concrete strength of core and cast specimens (ASTM 2018c).

Researchers have attempted to relate the compressive and splitting strengths of concrete cast cylinders with regression modeling (Hammitt 1974; Oloukun 1991; Thomas and Ramaswamy 2007; Pul 2008; Xu and Shi 2009; ACI 363R-10 2014; ACI 318-14 2014; Ramadoss 2014); Behnood et al. 2015), however, little literature was found relating the compressive and splitting strengths of concrete cores cylinders. Furthermore, no literature was found that examines the correlation between the compressive and splitting tensile strengths of both cast and core cylinders. Relationships between the compressive and splitting tensile strengths are desirable due to efficiency, similar strength testing procedures, and similar material mechanics. The present study uses the splitting tensile and compressive strengths of cast and core concrete specimens to develop mathematical regression models that can better correlate the splitting tensile and compressive strengths for high-strength concrete. The reliability of the proposed models is verified by comparing their predictive capabilities with the laboratory tests.

## 2 Background

### 2.1 Cylindrical Core Specimens

Determining the strength of an existing concrete pavement is crucial for maintenance and rehabilitation and is commonly achieved by extracting cylindrical core specimens from the pavement under consideration and then testing the core specimens in compression. Furthermore, core cylinders may also be necessary when the quality of the concrete during construction is in question (ASTM 2018c). The in-situ compressive strength, determined from testing cylindrical core specimens, is then used to assess the quality of the concrete (ASTM 2018c). Testing core cylinders in compression is attractive for QA/QC purposes due to the similarities between the testing procedures of core and cast cylindrical specimens. Furthermore, the tensile strength of core specimens is estimated in the same way as cast specimens by testing cores in splitting tension. The splitting tensile strength is determined by loading a core specimen in compression along its length until the specimen fails in indirect tension (splitting tension) due to internal tensile stresses.

### 2.2 Relationship Between Core and Cast Specimens

Generally, the compressive strength of core specimens is lower than that of cast specimens due to factors that potentially decrease the core strength, such as: core location and elevation, drilling orientation, drilling type, drill condition, temperature, moisture, and core damages (ASTM 2018c; Chen et al. 2013). A relationship between the two types of specimens (cast and core) would be ideal however, due to the aforementioned factors that decrease the strength of core cylinders, no such relationship exists between the two specimen types (ASTM 2018c). Instead of relating the two types of specimens, this study aims to combine both core and cast specimen strengths to better predict the compressive or splitting tensile strengths of either core or cast specimens.

### 2.3 Relationships Between Compressive and Splitting Strengths

Predicting the splitting strength with the compressive strength is efficient, cost effective, and straightforward. Researchers have related the concrete strengths of cast standard cylinders; however, little information was found on the relationships of core concrete strengths and no information was found on the relationships of combined core and cast concrete strengths. The mathematical model proposed by other researchers to relate the compressive and splitting tensile strengths are linear models and power models and are shown in Table 1. The nonlinear regression power model is the most common mathematical model used for relating the compressive and splitting strengths of concrete. This model type is also used to relate the compressive and flexural strengths of concrete. This particular model is attractive for relating concrete strengths due the nonlinear nature of concrete strength gain (Campos et al. 2020). Furthermore, because the compressive and splitting tensile strengths both fail by applying a compressive force to the cylinder, more credence is given to the strength relationship. The relationship between the compressive and splitting tensile strengths can be traced back to Raphael (1984) who proposed a nonlinear power regression model of the following form:

$$f_{st} = af_c^n \tag{1}$$

where, $f_{st}$ is the splitting tensile strength (indirect tensile), $a$ is a regression coefficient, $f_c$ is the compressive strength, and $n$ is a power regression coefficient.

**Table 1.** Relationship between splitting tensile and compressive strength

| Author/Code | Equation |
|---|---|
| Ramadoss (2014) | $f_{st} = 0.118f_c^{0.84}\ (MPa)$ |
| Xu and Shi (2009) | $f_{st} = 0.21f_c^{0.83}\ (MPa)$ |
| Oluokun (1991) | $f_{st} = 0.294f_c^{0.69}\ (MPa)$ |
| Hammitt (1974) | $f_{st} = 0.08f_c + 0.7014\ (MPa)$ |

*(continued)*

**Table 1.** (*continued*)

| Author/Code | Equation |
|---|---|
| Thomas and Ramaswamy (2007) | $f_{st} = 0.57\sqrt{f_c}\ (MPa)$ |
| Behnood et al. (2015) | $f_{st} = 0.219 f_c^{2/3}\ (MPa)$ |
| Pul (2008) | $f_{st} = 0.106 f_c^{0.948}\ (MPa)$ |
| ACI 363R-10 (2010) | $f_{st} = 0.59\sqrt{f_c}\ (MPa)$ |
| ACI 318-14 (2014) | $f_{st} = 0.56\sqrt{f_c}\ (MPa)$ |

Note: $f_{st}$ = *splitting tensile strength* $(MPa)$, $f_c$ = *compressive strength* $(MPa)$

# 3 Materials and Methods

## 3.1 Concrete Mixture Design

Cylindrical core and cast concrete specimens were obtained from two different airport pavement projects. The concrete mix proportions for the specimens were designed to meet the minimum required compressive strength of 34.5 MPa at 28 days, according to US Army Corps of Engineers EM 1110-2-2000 (U.S. Army Corps of Engineers 2001) design specifications. The aggregate size, aggregate moisture, and water to cement ratio were all adjusted until the mix designs met the specified strengths for both pavements at 28 days. The water to cement ratio for both mix designs is 0.43. Tables 2 and 3 show the mix design proportions for both the core and cast specimens. The concrete cast cylinders were fabricated, cured, and tested as per ASTM C31 (2018b), C39 (2018a), and C496 (2017) specifications and the core cylindrical specimens were extracted and tested as per ASTM C42 (2018c).

**Table 2.** Core concrete mix proportions

| Materials | (%) |
|---|---|
| Cement | 10.73 |
| Fly ash | 3.58 |
| Coarse aggregate | 49.78 |
| Fine aggregate | 30.69 |
| Water | 5.20 |
| Air entrainment | 0.01 |
| Water reducer | 0.02 |
| Set retarder | 0.00 |

**Table 3.** Cast concrete mix proportions

| Materials | (%) |
|---|---|
| Cement | 7.97 |
| Fly ash | 3.26 |
| Coarse aggregate | 44.37 |
| Fine aggregate | 24.50 |
| Water | 14.31 |
| Air entrainment | 0.02 |
| Water reducer | 0.06 |
| Air | 5.50 |

### 3.2 Test Methods

In total, 140 specimens were tested in this study. A portion of these specimens were tested under compression and another portion were tested in indirect tension (splitting tension). A total of 84 cylindrical core specimens each having a diameter of 101.6 mm. and a length of 203.2 mm. were extracted, prepared, and tested. Of these, 42 core cylinders were tested under uniaxial compressive loading as per ASTM C39 (2018a) and C42 (2018c) using a universal testing machine after curing for over 28 days. The remaining 42 core cylinders were tested under indirect tension after curing for over 28 days as per ASTM C496 (2017) and C42 (2018c) using a universal testing machine.

A total of 56 cylindrical cast specimens each having a diameter of 152.4 mm. and a length of 304.8 mm. were fabricated and tested. Of these, 28 cast cylinders were tested under uniaxial compressive loading as per ASTM C39 (2018a) using a universal testing machine after curing for 7 days. The remaining 28 cast cylinders were tested under indirect tension after curing for 7 days as per ASTM C496 (2018c) using a universal testing machine. The average compressive and splitting tensile stresses were then calculated for both the core and cast specimens. A comparison of the average core and cast stresses can be seen in Fig. 1. Due to concrete being mostly nonlinear with high variability, the r-squared values in Fig. 1. are ignored as they only represent the linear trendlines, which are provided to demonstrate the correlation direction. R-squared values are not valid for nonlinear models (Spiess and Neumeyer 2010). Therefore, the Root Mean Square Error (RMSE) will be used to assess the nonlinear regression models instead of the r-squared values.

### 3.3 Model Development

IBM SPSS Statistics 26, a statistical regression software, was used to analyze and determine mathematical relationships between the compressive and splitting tensile strengths of the concrete core and cast cylinders at 7 and 28 days. Generalized models were developed to predict the splitting tensile strength at potentially any day by simultaneously analyzing both 7-day cast and 28-day core concrete strengths. The compressive and

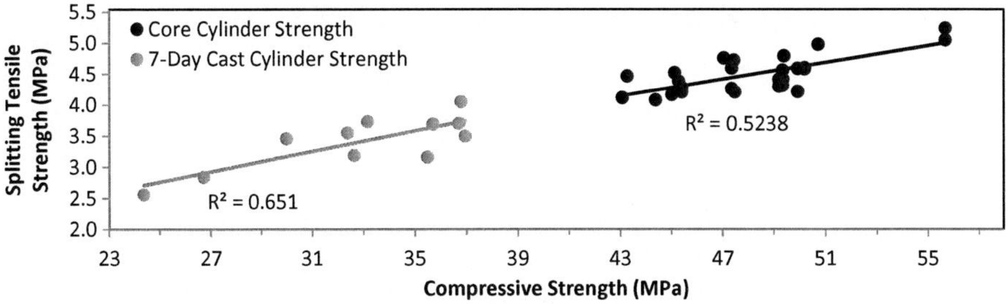


**Fig. 1.** Strength parameters of cast cylinders at 7 days and core cylinders over 28 days: splitting tensile strength vs. compressive strength.

splitting tensile strengths were related with the power, square root, log-linear, and linear regression models. The error in the prediction equations was assessed by analyzing the differences in the RMSE of each equation to determine the most accurate equation. The RMSE analysis was performed for each regression equation and specimen type (core and cast) to better understand the error between the two types of specimens.

# 4 Results and Discussion

## 4.1 Regression Models

The regression equations that were developed with core and cast concrete cylinder strengths are as follows:

$$f_{st} = 0.259 {f_c}^{0.735}\ (\text{MPa}) \tag{2}$$

$$f_{st} = 0.753 {f_c}^{0.532} - 1.458\ (\text{MPa}) \tag{3}$$

$$f_{st} = 0.094 f_c (\text{MPa}) \tag{4}$$

$$f_{st} = 0.116 \ln(f_c)^{2.695}\ (\text{MPa}) \tag{5}$$

$$f_{st} = 2.801 \ln(f_c) - 6.391\ (\text{MPa}) \tag{6}$$

$$f_{st} = 0.631\sqrt{f_c}\ (\text{MPa}) \tag{7}$$

where, $f_{st}$ is the splitting tensile strength (indirect tensile), ln is the natural log, and $f_c$ is the compressive strength.

## 4.2 Root Mean Square Error Analysis and Model Comparison

Results from the RMSE analysis can be seen in Tables 4 and 5. Table 4 shows the RMSE analysis for equations proposed by researchers/institutions and Table 5 shows the RMSE analysis for the regression equations developed in this study. From Table 5 it is observed that linear Eq. (4) and square root Eq. (7) have the largest RMSEs for cast and core specimens relative to the other regression equations. Consequentially, Eqs. (4) and (7) have the largest overall RMSEs. Log-linear Eq. (6) has the lowest RMSE for cast specimens (0.239 MPa) and a lower RMSE for core specimens (0.206 MPa). Power Eqs. (2), (3), and (5) have nearly identical RMSEs for cast and core specimens. The power equations gave consistent and low RMSEs for each case. The complexity of the power model had almost no effect on the RMSE for each case. From Fig. 2 it is observed that proposed power Eq. (2) and Eq. (5) both trend well with the concrete strengths of cast and core specimens.

**Table 4.** Root mean square error analysis

| Author/Code | RMSE cast 7 day | RMSE core 28 day | Overall RMSE |
|---|---|---|---|
| Ramadoss (2014) | 0.294 | 0.464 | 0.453 |
| Xu and Shi (2009) | 0.505 | 0.807 | 0.785 |
| Oluokun (1991) | 0.266 | 0.282 | 0.298 |
| Hammitt (1974) | 0.247 | 0.224 | 0.248 |
| Thomas and Ramaswamy (2007) | 0.291 | 0.546 | 0.521 |
| Behnood et al. (2015) | 0.440 | 0.496 | 0.516 |
| Pul (2008) | 0.523 | 0.349 | 0.439 |
| ACI 363R-10 (2010) | 0.268 | 0.422 | 0.412 |
| ACI 318-14 (2014) | 0.317 | 0.611 | 0.581 |

RMSE = Root mean square error (MPa)

**Table 5.** Regression equation error analysis

| Equation | RMSE Cast 7 Day | RMSE Core 28 Day | Overall RMSE |
|---|---|---|---|
| Regression Eq. 2 | 0.242 | 0.202 | 0.231 |
| Regression Eq. 3 | 0.240 | 0.202 | 0.231 |
| Regression Eq. 4 | 0.373 | 0.229 | 0.300 |
| Regression Eq. 5 | 0.240 | 0.203 | 0.231 |
| Regression Eq. 6 | 0.239 | 0.206 | 0.233 |
| Regression Eq. 7 | 0.358 | 0.227 | 0.293 |

RMSE = Root mean square error (MPa)

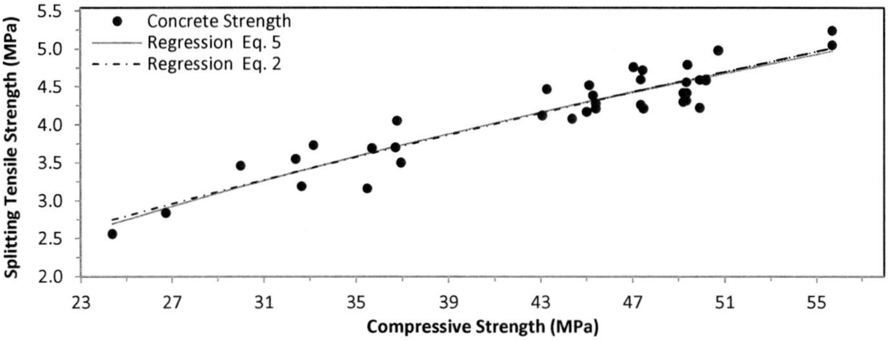


**Fig. 2.** Proposed splitting tensile strength prediction model capabilities: splitting tensile strength vs. compressive strength.

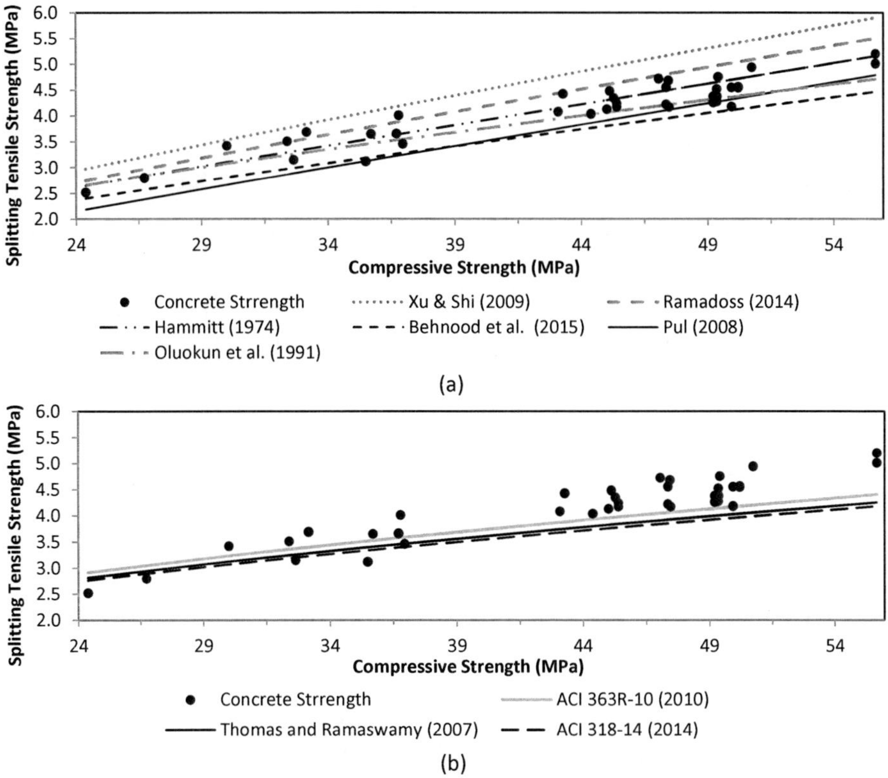


**Fig. 3.** Splitting tensile strength model prediction capabilities by other researchers: splitting tensile strength vs. compressive strength.

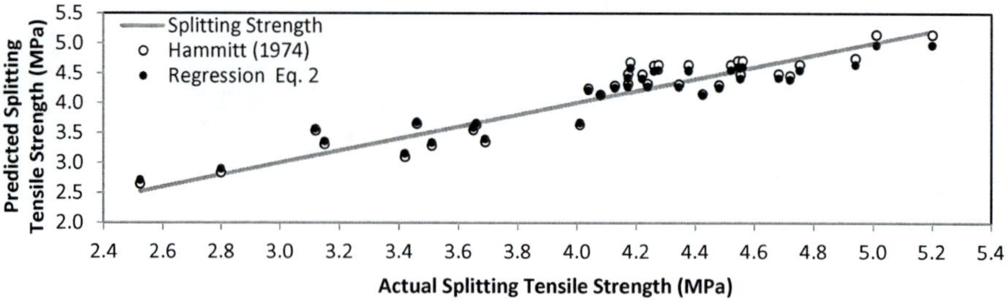


(a)

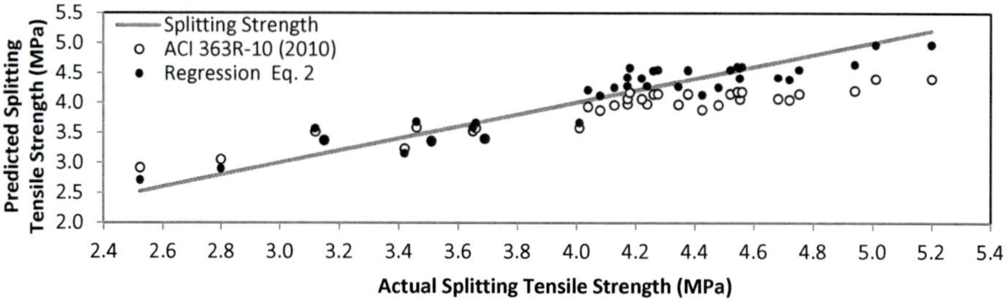


(b)

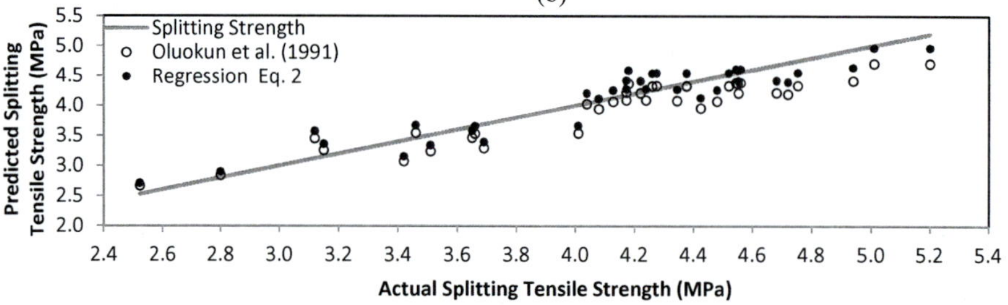


(c)

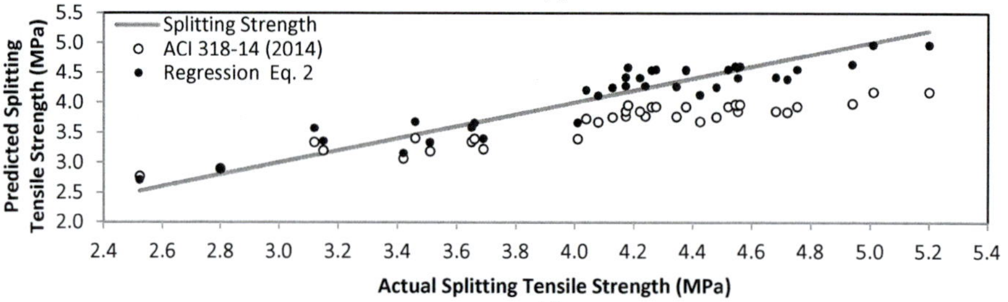


(d)

**Fig. 4.** Predicted splitting tensile strength vs. actual splitting tensile strength: (a) comparison of splitting tensile strength model by Hammitt (1974) and proposed Eq. (2), (b) comparison of splitting tensile strength model by ACI 363R-10 (2010) and proposed Eq. (2), (c) comparison of splitting tensile strength model by Oluokun (1991) and proposed Eq. (2), comparison of splitting tensile strength model by ACI 318-14 (2014) and Proposed Eq. (2).

From Table 4 it is observed that the equations from the literature generally have lower RMSEs for cast specimens than core specimens. This is most likely due to the exclusion of core specimens in the regression modeling. Furthermore, the overall RMSEs for the equations from the literature are all relatively high. This is most likely due to the lack of core specimens and small sample sizes. Figure 3 shows that the square root model is inadequate for correlating the compressive and splitting tensile strengths. The underperforming predictive capabilities of the square root model can be seen more clearly in Fig. 4. Figure 4 also shows that a linear equation with a regression constant increases its predictive capabilities. Comparatively, proposed Eq. (2) has a lower RMSE than all of the equations from the literature in each case.

## 5 Conclusions

The present study used the splitting tensile (indirect tensile) and compressive strengths of cast and core specimens to develop mathematical regression models that can better correlate the indirect splitting tensile and compressive strengths at 28 days for high-strength concrete. The reliability of the proposed models was verified by comparing their predictive capabilities with the measured results. Statistical regression software was used to analyze the cast and core strength test results and develop regression models. From the root mean square error and regression model analysis, the following conclusions can be made:

- Regression equations that only utilize the strength parameters from cast specimens leads to larger errors when predicting the strengths of core specimens. Both core and cast specimens should be analyzed together to produce regression equations that can predict either core or cast strengths.
- The nonlinear power model was the most accurate model developed for relating the compressive and splitting tensile strengths for both core and cast specimens. Linear and square root models are not adequate for relating the compressive and splitting tensile strengths. However, including an additional regression constant to the linear model improves its predictive capabilities.
- The compressive strength has a positive nonlinear correlation with the splitting tensile strength for both core and cast specimens. This is likely due to similar specimen geometry and testing methods.
- The proposed model is adequate for evaluating the splitting tensile strength of an existing pavement. Further, the proposed equation is able to predict either the cast or core splitting strengths allowing designers to determine the strength of new and existing pavements. The proposed power model can be used to predict the splitting tensile strength at potentially any day.

**Acknowledgements.** The funding for this project was provided by the Oklahoma Center for the Advancement of Science and Technology (OCAST) and ROCA Engineering, Inc.

## References

ACI (American Concrete Institute): Building code requirements for structural concrete and commentary. ACI 318-2014. ACI, Farmington Hills (2014)

ACI Committee 363 Report on high-strength concrete. ACI 363R-10, American Concrete Institute Committee 363. ACI, Farmington Hills (2010)

ASCE (American Society of Civil Engineers): 2017 Infrastructure Report Card. ASCE's 2017 Infrastructure Report Card (2017). https://www.infrastructurereportcard.org/cat-item/roads/. Accessed 25 July 2020

ASTM (American Society for Testing and Materials): Standard test method for splitting tensile strength of cylindrical concrete specimens. C496/C496M – 17, West Conshohocken, PA (2017)

ASTM (American Society for Testing and Materials): Standard test method for compressive strength of cylindrical concrete specimens. C39/C39M – 18, West Conshohocken, PA (2018a)

ASTM (American Society for Testing and Materials): Standard test method for Making and Curing Concrete Test Specimens in the Field. C31/C31M – 18b, West Conshohocken, PA (2018b)

ASTM (American Society for Testing and Materials): Standard test method for obtaining and testing drilled cores and sawed beams of concrete. C42/C42M – 18, West Conshohocken, PA (2018c)

Behnood, A., Verian, K.P., Gharehveran, M.M.: Evaluation of the splitting tensile strength in plain and steel fiber-reinforced concrete based on the compressive strength. Constr. Build. Mater. **98**, 519–529 (2015)

Campos, R., Mendez Larrain, M.M., Zaman, M., Pozadas, V.: Relationships between compressive and flexural strengths of concrete based on fresh field properties. Int. J. Pavement Res. Technol. (2020). https://doi.org/10.1007/s42947-020-1074-0

Chen, D.H., Won, M.: Field investigations of cracking on concrete pavements. J. Perform. Constr. Facil. **21**(6), 450–458 (2007)

Chen, X., Wu, S., Zhou, J.: Analysis of mechanical properties of concrete cores using statistical approach. Mag. Concrete Res. **65**(24) (2013). https://doi.org/10.1680/macr.13.00113

Hammitt, G.M.: Concrete strength relationships, U.S. Vicksburg, Miss: Army Engineer Waterways Experiment Station, Soils and Pavements Laboratory (1974)

Hoła, J., Bień, J., Sadowski, L., Schabowicz, K.: Non-destructive and semi-destructive diagnostics of concrete structures in assessment of their durability. Bull. Pol. Acad. Sci. Tech. Sci. **63**, 87–96 (2015)

Justo-Silva, R., Ferreira, A.: Pavement maintenance considering traffic accident costs. Int. J. Pavement Res. Technol. **12**(6), 562–573 (2019). https://doi.org/10.1007/s42947-019-0067-3

Oloukun, A.F.: Prediction of concrete tensile strength from its compressive strength: evaluation of existing relations for normal weight concrete. ACI Mater. J. **88**(3), 302–309 (1991)

Pul, S.: Experimental investigation of tensile behavior of high strength concrete. Indian J. Eng. Material Sci. **15**(6), 467–472 (2008)

Ramadoss, P.: Correlation of compressive strength and other engineering properties of high-performance steel fiber–reinforced concrete. ASCE J. Mater. Civ. Eng. **27**(1), 04014114 (2014). https://doi.org/10.1061/(ASCE)MT.1943-5533.0001050

Raphael, J.M.: Tensile strength of concrete. ACI J. **81**(2), 158–165 (1984)

Spiess, A.N., Neumeyer, N.: An evaluation of R2 as an inadequate measure for nonlinear models in pharmacological and biochemical research: a Monte Carlo approach. BMC Pharmacol. **10**(1), 6 (2010). https://doi.org/10.1186/1471-2210-10-6

Thomas, J., Ramaswamy, A.: Mechanical properties of steel fiber reinforced concrete. ASCE J. Mater. Civ. Eng. **19**(5), 385–392 (2007)

U.S. Army Corps of Engineers: Standard Practice for Concrete for Civil Works Structures Proponent. CECW-EG. EM 1110-2-2000. USACE, Washington DC, USA (2001)

U.S. Department of Transportation: "2015 Status of the Nation's Highways, Bridges and Transit: Conditions and Performance." Federal Highway Administration/ Department of Transportation (2015). https://www.fhwa.dot.gov/policy/2015cpr/pdfs.cfm. Accessed 25 July 2020

Xu, B.W., Shi, H.S.: Correlations among mechanical properties of steel fiber reinforced concrete. Const. Build. Mater. **23**(12), 3468–3474 (2009). https://doi.org/10.1016/j.conbuildmat.2009.08.017

Yoder, E.J., Witczak, M.W.: Principles of Pavement Design, 2nd edn., p. 81. Wiley (1975). https://app.knovel.com/hotlink/toc/id:kpPPDE000N/principles-pavement-design/principles-pavement-design

# Predicting the Small Strain Stiffness of a Calcareous Sand Considering Sample Preparation Method and Stress Path

Jinquan Shi[1(✉)] and Wim Haegeman[2]

[1] School of Civil Engineering, Chongqing University, Chongqing 400045, China
jinquan.shi@cqu.edu.cn

[2] Department of Civil Engineering, Ghent University, Ghent, Belgium

**Abstract.** This study examines the effect of the sample preparation method on the empirical-determined constants in the Hardin equation for predicting the small strain stiffness of a calcareous sand. Multi-directional bender element tests are performed in a triaxial cell under isotropic and anisotropic consolidation stress paths. Five sample preparation methods as air and water pluviation, dry and moist tamping and dry funnel deposition are used as the sample reconstitution techniques. Test results show that the sample preparation method has a considerable impact on the constants in the Hardin equation. Moist tamping samples show the highest A and the lowest n compared with the samples made by the other preparation methods. The lowest A and the highest n are found in the dry funnel deposition method. It is concluded that a higher A always corresponds with a lower n and the magnitude of the constants are related to the stability of the sample fabric. It is also shown that the constants determined in the anisotropic consolidation tests correspond well with the constants in the isotropic consolidation.

## 1 Introduction

Small strain shear modulus, $G_0$, is a fundamental parameter in determining the ground movements of many geotechnical infrastructures. Soils like sands and clay behave elastic at the very small strain ranges ($\leq 10^{-5}\%$) in which $G_0$ can be determined by some indirect measurement techniques, for example, bender element test (Bellotti et al. 1996; Pennington et al. 1997; Fioravante 2000; Escribano and Nash 2015; Shi et al. 2020a). Based on the shear wave velocity measurement in bender element test, $G_0$ can be evaluated with Eq. (1) which is expressed as

$$G_0 = \rho V_s^2 \tag{1}$$

where $G_0$ is the initial shear modulus (MPa); ρ is the bulk density (kg/m$^3$) and $V_s$ is the shear wave velocity (m/s).

It has been proved that the small strain stiffness of sands is affected by factors such as physical properties of sands (coefficient of uniformity $C_u$, particle size, particle shape


H. Khabbaz et al. (Eds.): GeoChina 2021, SUCI, pp. 70–77, 2021.
https://doi.org/10.1007/978-3-030-79650-1_6

and fine content), soil fabric, stress path and stress history. Hardin and Black (1966) proposed an equation for predicting $G_0$ as

$$G_0 = Af(e)(\frac{\sigma'}{P_a})^n \tag{2}$$

where $A$ and $n$ are experimentally determined constants; $\sigma'$ is the mean effective stress; $P_a$ is the reference pressure (≈the atmospheric 100 kPa); $f(e)$ is the void ratio function. Studies as Iwasaki and Tatsuoka (1977), Wichtmann and Triantafyllidis (2009, 2010), Payan et al. (2017), Giang et al. (2017) and Jafarian and Javdanian (2020) discussed and quantified the factors affecting $G_0$ and proposed the modified expressions of the Hardin equation. Referring to these literatures, the constants in Eq. (2) can be expressed as the functions of the quantified factors which are $C_u$, mean grain size $D_{50}$, fine content $FC$, stress history OCR and particle regularity $\rho$. However, the constants in these studies are obtained from the samples prepared by single method and the effect of sample preparation technique on the constants is rarely discussed.

In this study, five methods as air and water pluviation, dry and moist tamping and dry funnel deposition are used for preparing samples with a calcareous sand from Persian Gulf. Shear wave velocities are measured in both vertical and horizontal directions in the bender element tests under isotropic consolidation. In addition, in order to investigate the effect of stress path on the constants in Eq. (2), two series of tests are performed under anisotropic consolidation stress path. Finally, the effects of sample preparation method and stress path on the constants in the Hardin equation are evaluated.

## 2 Test Materials

The calcareous sand, named S1 sand, is from a reclamation site in Persian Gulf. It is mainly composed of marine deposits like shell and coral so the $CaCO_3$ content of the sand can reach over 95%. The physical properties of S1 sands are: specific gravity $G_s$ = 2.81; $C_u = 3.89$; $D_{50} = 0.602$ mm; $e_{max} = 1.19$; $e_{min} = 0.73$. The particle size gradation of S1 sands is shown in Fig. 1.

## 3 Test Program

Five methods as air and water pluviation, dry and moist tamping and dry funnel deposition method are used to reconstitute samples with 50 mm and 100 mm in diameter and height respectively. All samples are prepared in a spilt mold. The air and water pluviation methods are carried out with the specialized designed pluviators. Five-layer sieves are arranged in the pluviators so the sands falling from the funnels can be uniformly distributed. The density of the air pluviation sample is adjusted by controlling the opening size of the funnel. The densification of the water pluviation sample depends on tapping the mold side. The dry and moist tamping samples are prepared with the undercompaction method proposed by Ladd (1978). Five and ten compacted layers are used in the dry and moist tamping samples respectively. Sands with 5% water content by mass are prepared for moist tamping specimen. The dry funnel deposition sample

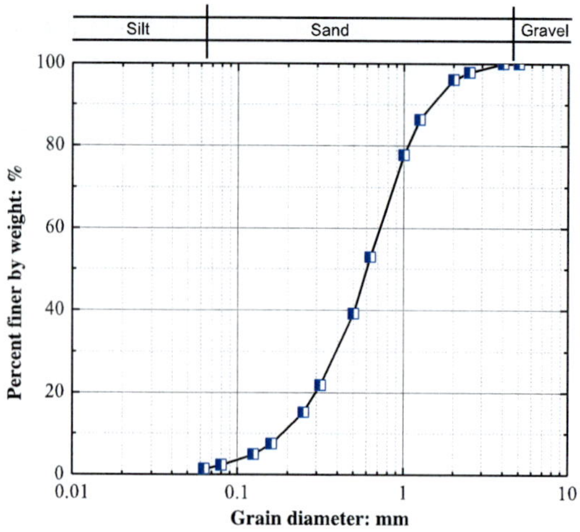


**Fig. 1.** Particle size gradation of S1 sands

is prepared with a long neck funnel in which the opening is almost zero to the sand surface during deposition. The densification is carried out by tapping the mold side with a rubber hammer. Details of the five methods can be found in Shi et al. (2019), Shi et al. (2020a) and Shi et al. (2020b). All the tests are performed at least two times to verify the repeatability.

All the samples are saturated with the succession use of carbon dioxide, deaired water and back pressure. After saturation, samples are consolidated under the stress path shown in Fig. 2. The bender element tests are performed at the end of each loading stage. Single since waves are excited at the frequency of 15 kHz and with the amplitude of 20 V. Peak to peak method is used for signal interpretation.

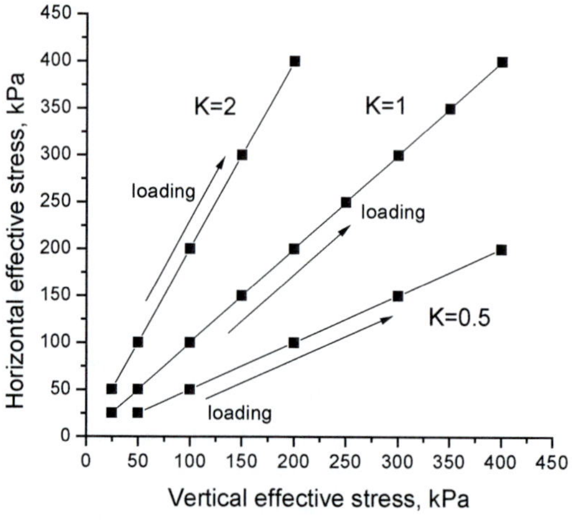


**Fig. 2.** Isotropic and anisotropic consolidation stress path

# 4 Test Results

The void function $f(e)$ by Jamiolkowski et al. (1995) is used in this study, which is expressed as

$$f(e) = e^{-d} \tag{3}$$

Figure 3 shows that $G_0$ of the samples prepared by different methods versus void ratios under the isotropic effective consolidation stress of 100 kPa. $G_{HH}$, $G_{HV}$ and $G_{VH}$ are three shear modulus determined by the shear wave velocities measured on both the vertical and horizontal planes and the first and second subscript denote the direction of particle propagation and polarization respectively. A unified trend between the three modulus and void ratio can be found in Fig. 3. Finally, an averaged curve is plotted and a $d$ value of 1.195 in Eq. (4) is obtained. This value is quite approximate to the results of 1.2 and 1.3 in Fioravante et al. (2013) and Fioravante et al. (1998) for Kenya and Quiou calcareous sands. Therefore, with $d$ of 1.195, $G_0$ is normalized by $f(e)$ for the regression analysis of the coefficients $A$ and $n$, as shown in Fig. 4. The coefficients $A$ and $n$ from the samples prepared by the five methods are summarized in Table 1. It is seen that some differences are shown among the samples. It is generally recognized that $A$ represents soil type, particle characteristic and fabric. $n$ denotes the response of soils to stress variation. From Table 1, the higher $A$ corresponds to the higher shear modulus reported by Shi et al. (2020a). It can also be seen that the samples with a higher $A$ always show a lower $n$. This also verifies the deduction mentioned above since a smaller $n$ means the less strain response of the sample subjected to loading and the stronger structure the sample has.

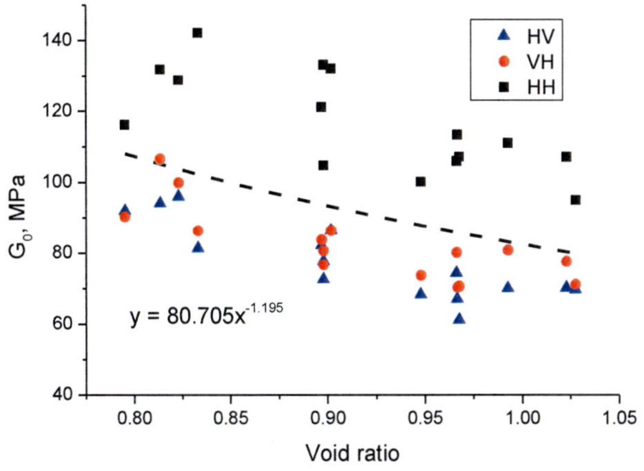


**Fig. 3.** $G_0$ versus void ratio at the effective consolidation stress of 100 kPa.

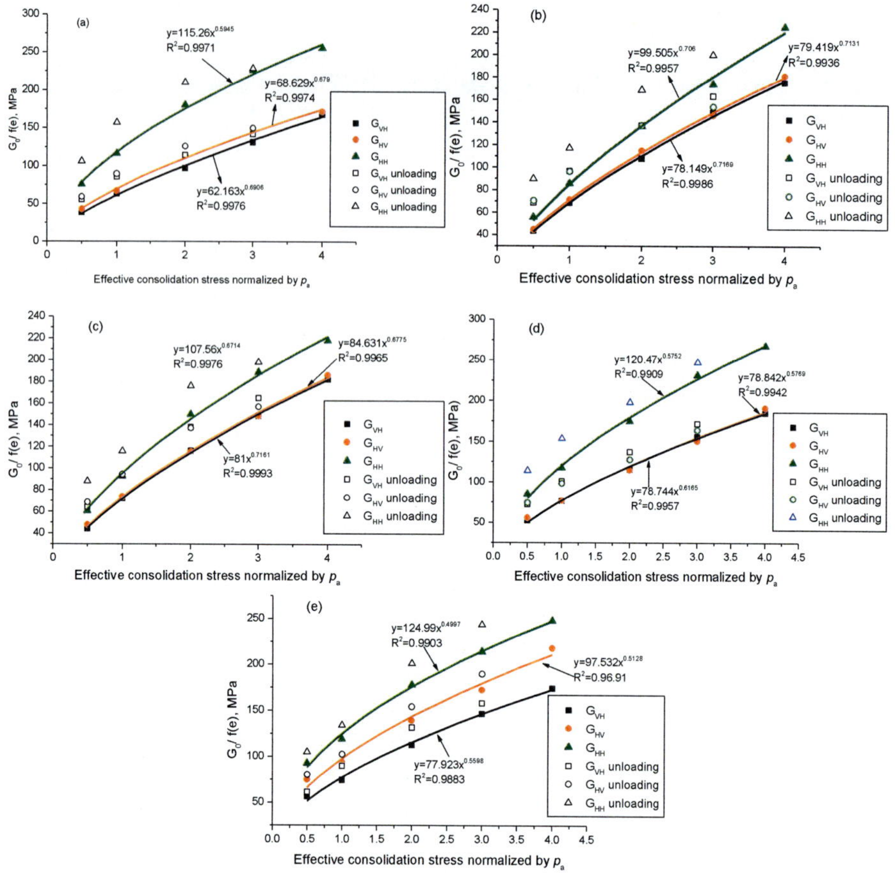


**Fig. 4.** $G_0/f(e)$ versus the normalized effective stress: (a) air pluviation; (b) dry funnel deposition; (c) dry tamping; (d) water pluviation; (e) moist tamping.

**Table 1.** Summary of the coefficients $A$ and $n$

| Shear modulus | Coefficients | Air pluviation | Dry funnel deposition | Dry tamping | Water pluviation | Moist tamping |
|---|---|---|---|---|---|---|
| $G_{HH}$ | $A$ | 115.26 | 87.334 | 94.212 | 124.47 | 124.99 |
| | $n$ | 0.5945 | 0.6695 | 0.6367 | 0.5752 | 0.4997 |
| $G_{HV}$ | $A$ | 68.629 | 69.705 | 74.126 | 78.842 | 97.532 |
| | $n$ | 0.679 | 0.6766 | 0.6427 | 0.5769 | 0.5128 |
| $G_{VH}$ | $A$ | 62.163 | 68.59 | 70.945 | 78.744 | 77.923 |
| | $n$ | 0.6906 | 0.6804 | 0.6814 | 0.6165 | 0.5598 |

To take into account the anisotropic effective stress state, Eq. (2) is modified as

$$G_0 = Af(e)(\frac{\sigma_h'}{P_a})^{n_h}(\frac{\sigma_v'}{P_a})^{n_v} \tag{4}$$

where $\sigma_h'$ and $\sigma_v'$ are the effective horizontal and vertical stresses; $n_h$ and $\sigma_v'$ are the experimental determined constants. For the three shear moduli $G_{HH}$, $G_{HV}$ and $G_{VH}$, $\sigma_h'$ and $\sigma_v'$ are denoted as the principle effective stress along the directions of wave propagation or polarization. For example, $G_{HH}$ is determined by the shear wave propagating and polarizing in the horizontal direction. Therefore, the vertical stress $\sigma_v'$ has little impact on $G_{HH}$.

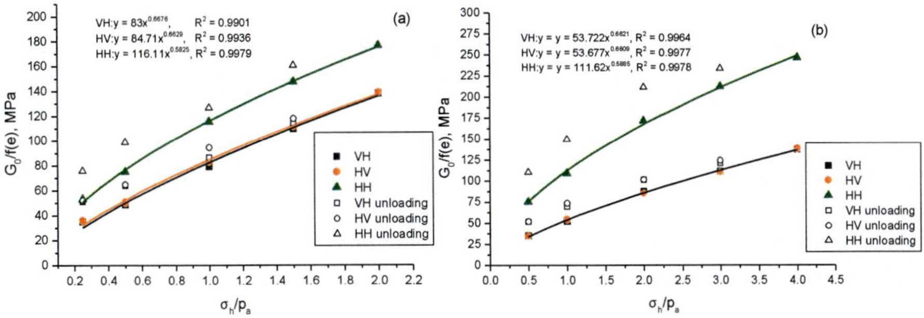


**Fig. 5.** $G_0/f(e)$ versus normalized horizontal effective stress: (a) K = 0.5; (b) K = 2.

To address the effect of the anisotropic stress state on the small stain stiffness of S1 sands, tests under anisotropic consolidation with K = 0.5 and K = 2 are performed, as shown in Fig. 5. The effect of sample preparation method is not considered under the anisotropic stress state and only the samples prepared by the air pluviation method are tested. It is assumed that the constants $A$ and void ratio function are identical with those established in the isotropic consolidation tests. It is necessary to mention that based on the cross anisotropic theory, $A$ of $G_{HV}$ and $G_{VH}$ should be identical. The slight difference obtained in this study is deemed as the random error. Therefore, for the anisotropic consolidation tests, $A$ is averaged from the values of $G_{HV}$ and $G_{VH}$. Figure 5 plots $G_0/f(e)$ versus horizontal effective stress $\sigma_h'$. It is known that $\sigma_h' = 0.5\,\sigma_v'$ and $\sigma_h' = 2\,\sigma_v'$ for K = 0.5 and K = 2 tests respectively. Therefore, Eq. (4) can be modified as

$$\frac{G_0}{f(e)} = A(\frac{\sigma_h'}{P_a})^{n_h+n_v} \cdot (\frac{1}{2})^{n_v} \quad \text{when K} = 0.5 \tag{5}$$

$$\frac{G_0}{f(e)} = A(\frac{\sigma_h'}{P_a})^{n_h+n_v} \cdot (2)^{n_v} \quad \text{when K} = 0.5 \tag{6}$$

The constants $n_h$, $n_v$ evaluated from K = 0.5 and K = 2 tests are summarized in Table 2. It is seen the constants obtained from the two anisotropic consolidation tests are approximately identical and the sum of $n_h$ and $n_v$ is also approximate to the $n$ evaluated in the isotropic consolidation tests, meaning that the constants for predicting the small strain stiffness of S1 sands are not influenced by the anisotropic stress state.

**Table 2.** Constants $n_h$, $n_v$ for the anisotropic consolidation tests.

| K | Constants | HH | HV | VH |
|---|---|---|---|---|
| 0.5 | $n_h$ | 0.522 | 0.307 | 0.307 |
| | $n_v$ | 0.061 | 0.358 | 0.358 |
| 2 | $n_h$ | 0.543 | 0.321 | 0.321 |
| | $n_v$ | 0.046 | 0.340 | 0.340 |

## 5 Conclusions

In this study, the bender element tests are performed on calcareous sand samples to investigate the effects of sample preparation method and stress path on the constants in the Hardin equation for predicting small strain modulus $G_0$. Five methods as air and water pluviation, dry and moist tamping and dry funnel deposition are used as the sample reconstitution techniques. It is found that the samples prepared by different methods show differences in the constants *A* and *n*, behaving that the samples with stronger fabric show higher A and lower *n*. The constants summarized from isotropic and anisotropic consolidation tests match well.

## References

Bellotti, R., Jamiolkowski, M., Lo Presti, D.C.F., O'Neill, D.A.: Anisotropy of small strain stiffness in Ticino sand. Géotechnique **46**(1), 115–131 (1996). https://doi.org/10.1680/geot.1996.46.1.115

Escribano, D.E., Nash, D.F.T.: Changing anisotropy of $G_0$ in Hostun sand during drained monotonic and cyclic loading. Soils Found. **55**(5), 974–984 (2015)

Ezaoui, A., Benedetto, H.D.: Experimental measurements of the global anisotropic elastic behaviour of dry Hostun sand during triaxial tests, and effect of sample preparation. Géotechnique **59**(7), 621–635 (2009)

Fioravante, V.: Anisotropy of small strain stiffness of Ticino and Kenya Sands from seismic wave propagation measured in triaxial testing. Soils Found. **40**(4), 129–142 (2000). https://doi.org/10.3208/sandf.40.4_129

Fioravante, V., Jamiolkowski, M., Ghionna, V.N., Pedroni, S.: Stiffness of carbonateQuiou sand from CPT. In: Roberston, M. (ed.) Proceeding of the First International Conference on Site Characterisation, pp. 1039–1049 (1998)

Fioravante, V., Giretti, D., Jamiolkowski, M.: Small strain stiffness of carbonate Kenya Sand. Eng. Geol. **161**, 65–80 (2013). https://doi.org/10.1016/j.enggeo.2013.04.006

Giang, P.H.H., Van Impe, P.O., Van Impe, W.F., Menge, P., Haegeman, W.: Small-strain shear modulus of calcareous sand and its dependence on particle characteristics and gradation. Soil. Dyn. Earthq. Eng. **100**, 371–379 (2017)

Hardin, B.O., Black, W.L.: Sand stiffness under various triaxial stresses. J. Soil Mech. Found. Div. **92**(2), 27–42 (1966)

Iwasaki, T., Tatsuoka, F.: Effects of grain size and grading on dynamic shear moduli of sands. Soils Found. **17**(3), 19–35 (1977)

Jafarian, Y., Javdanian, H.: Dynamic properties of calcareous sand from the Persian Gulf in comparison with siliceous sands database. Int. J. Civ. Eng. **18**(2), 245–249 (2019). https://doi.org/10.1007/s40999-019-00402-9

Ladd, R.S.: Preparing test specimens using undercompaction. Geotech. Test. J. **1**(1), 16–23 (1978)

Payan, M., Senetakis, K., Khoshghalb, A., Khalili, N.: Characterization of the small-strain dynamic behaviour of silty sands; contribution of silica non-plastic fines content. Soil. Dyn. Earthq. Eng. **102**, 232–240 (2017)

Pennington, D.S., Nash, D.F.T., Lings, M.L.: Anisotropy of $G_0$ shear stiffness in Gault Clay. Géotechnique **47**(3), 391–398 (1997)

Shi, J., Haegeman, W., Mascini, A., Cnudde, V.: X-ray analysis on the effect of sample preparation on the microstructure of calcareous sands. Mar. Georesour. Geotech. **39**(3), 302–311 (2019)

Shi, J., Haegeman, W., Cnudde, V.: Anisotropic small strain stiffness of calcareous sand affected by sample preparation, particle characteristic and gradation. Géotechnique (2020a). https://doi.org/10.1680/jgeot.18.P.348

Shi, J., Haegeman, W., Andries, J.: Investigation on the mechanical properties of a calcareous sand: the role of the initial fabric. Mar. Georesour. Geotech. 1–17 (2020b)

Wichtmann, T., Triantafyllidis, T.: Influence of the grain-size distribution curve of quartz sand on the small strain shear modulus $G_{max}$. J. Geotech. Geoenviron. Eng. **135**(10), 1404–1418 (2009). https://doi.org/10.1061/(ASCE)GT.1943-5606.0000096

Wichtmann, T., Triantafyllidis, T.: On the influence of the grain size distribution curve on P-wave velocity, constrained elastic modulus Mmax and Poisson's ratio of quartz sands. Soil. Dyn. Earthq. Eng. **30**(8), 757–766 (2010). https://doi.org/10.1016/j.soildyn.2010.03.006

# Explicit Integration and Implementation of State-Dependent Constitutive Model for Rockfill Materials

Zengchun Sun[1], Hao Cui[1], Hanlong Liu[1,2], Chenggui Wang[1], Yang Xiao[1,2], and Huanran Wu[1,2(✉)]

[1] School of Civil Engineering, Chongqing University, Chongqing 400045, China
[2] Key Laboratory of New Technology for Construction of Cities in Mountain Area, Chongqing University, Chongqing 400045, China

**Abstract.** The dilatancy and stress-strain relationship of rockfill materials often depend on their material states. To better simulate the strength and deformation behaviors of rockfill materials under complex stress states, the state void-ratio index was incorporated into Rowe's stress-dilatancy equation, bounding stress ratio, and plastic modulus. A state-dependent constitutive model was established within the framework of the critical state theory and the bounding surface plasticity theory. The proposed model can well describe the strain hardening, strain softening, and dilatancy behaviors of rockfill materials. Thereafter, the adaptive sub-stepping explicit integration algorithm was adopted in the state-dependent model and implemented into a finite element code. To validate the proposed scheme, a numerical model was established at element scale to simulate drained triaxial compression tests in a large density and pressure ranges. Comparisons between the simulation results and experimental data, the performance of the explicit integration algorithm for the state-dependent constitutive model are analyzed. The accuracy and efficiency of the explicit integration to describe the state-dependent behavior of rockfill materials is verified, which lays a foundation for further engineering applications.

**Keywords:** Rockfill material · State-dependent · Explicit integration · Adaptive sub-stepping · Numerical implementation

## 1 Introduction

Rockfill, a kind of granular materials, is commonly applied in dam construction and railway subgrade engineering given its high strength, high hydraulic permeability and low cost. Great efforts have been devoted to investigate the shear behavior of rockfill materials due to its significance to the safety and stability of dams and embankments. Experimental results (Salim and Indraratna 2004; Xiao et al. 2014a) revealed that rockfill materials usually exhibit strain softening (or hardening) and volumetric expansion (or contraction) under complex loading.


H. Khabbaz et al. (Eds.): GeoChina 2021, SUCI, pp. 78–93, 2021.
https://doi.org/10.1007/978-3-030-79650-1_7

The hyperbolic elastic model (Duncan and Chang 1970) was commonly adopted to describe the nonlinear stress-strain behaviors of rockfill materials, but it cannot capture strain softening and dilatancy. To describe the complex mechanical behaviors of rockfill materials, many new or modified constitutive models were established. Varadarajan et al. (2006) presented a constitutive model to describe the stress-strain behavior of rockfill materials based on the disturbed state concept. Xiao et al. (2014a) established a new bounding surface plasticity model, considering the dependence of dilatancy and hardening on state parameters. Combining the drop-shaped yield surface, Yao et al. (2018) proposed a unified hardening model for granular soils in a large stress range by incorporating the limit void ratio in the double logarithmic scale and state parameter. Yin et al. (2017) developed an elastoplastic model for rockfill materials within the framework of critical state double-yield surface theory, which considers both particle breakage and size effect.

Numerical integration plays a pivotal role in solving practical boundary value problems. Accurate and effective integration of the constitutive equations at the stress point level first is essential to model rockfill materials within the finite element framework. Sloan (2001) introduced a sub-stepping explicit integration which automatically divide the strain increment into subincrements based on an estimation of the local error. This scheme can significantly enhance the accuracy and efficiency. Sheng et al. (2003) applied the adaptive sub-stepping explicit integration to the elastoplastic constitutive model of unsaturated soils. Hu and Liu (2014) established implicit and adaptive sub-stepping explicit integration for the anisotropic boundary surface model of saturated clay under cyclic loading. Combined with a novel loading-unloading decision method, Zhou and Zhang (2015) extended the adaptive sub-stepping explicit integration to the non-isothermal unified hardening model. However, the adaptive sub-stepping explicit integration has not been applied to the complex constitutive models for rockfill materials, and the models are not widely adopted in finite element software.

The main objective of this paper is to establish a state-dependent constitutive model by introducing state void-ratio index into Rowe's stress-dilatancy equation within the critical state framework. An adaptive sub-stepping explicit integration algorithm is adopted for the state-dependent model and implemented into a finite element code. The performance of the state-dependent model and the explicit scheme are investigated at element scale by simulating drained triaxial compression tests at various densities and pressures.

## 2 State-Dependent Constitutive Model

### 2.1 State Parameter

The stress-strain behaviors of soils, such as dilatancy, strain-hardening and strain-softening are related to the state of soils. The state parameter is of great significance to establish a unified constitutive model, e.g., the state parameter $\psi$ proposed by Been and Jefferies (1985), the state void-ratio index $I_e$ proposed by Wan and Guo (1998), the state pressure index $I_p$ proposed by Wang et al. (2002), and the state void-ratio-pressure index $I_{ep}$ proposed by Xiao and Liu (2017). The state void-ratio index $I_e$ was adopted to describe the state-dependent behavior of rockfill materials, which can be expressed as:

$$I_e = \frac{e}{e_{cs}} \tag{1}$$

where $e$ is the current void ratio; and $e_{cs}$ is the critical state void ratio.

The linear function of the critical state line (CSL) in the $e - (p'/p_a)^{\xi}$ plane can be expressed as:

$$e_{cs} = e_{\Gamma} - \lambda_{cs}(p'_{cs}/p_a)^{\xi} \tag{2}$$

where $p'_{cs}$ is the critical state mean effective stress; $p_a$ is the atmospheric pressure; $e_{\Gamma}$ is the initial critical state void ratio; $\lambda_{cs}$ is the slope of CSL; and $\xi$ is a material parameter. The definition of the state void-ratio index in the $e - (p'/p_a)^{\xi}$ plane is displayed in Fig. 1.

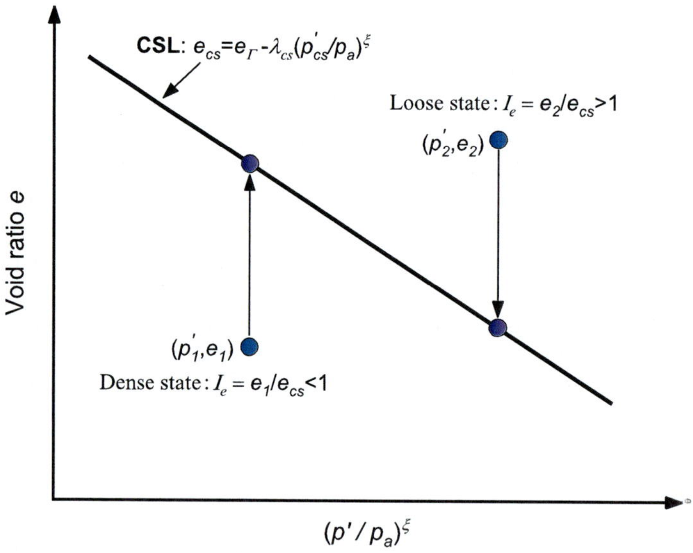


**Fig. 1.** Definitions of the state void-ratio index

## 2.2 Yielding Surface

Following the work by Li and Dafalias (2000), a simple yielding surface is adopted in this study:

$$f = q - \eta p' = 0 \tag{3}$$

where $\eta$ is the stress ratio.

The mean effective stress $p'$ and deviatoric stress $q$ are defined as:

$$p' = I_1/3 \tag{4a}$$

$$q = \sqrt{3J_2} \tag{4b}$$

where $I_1$ is the first invariant of the effective stress tensor; and $J_2$ is the second invariant of the deviatoric stress tensor.

The loading unit normal vector $\mathbf{n}_f = \left[n_p^f, n_q^f\right]^T$ are given as:

$$n_{p'}^f = \frac{\frac{\partial f}{\partial p'}}{\left\| \frac{\partial f}{\partial \underset{\sim}{\sigma}} \right\|} = \frac{\frac{\partial f}{\partial p'}}{\sqrt{\left(\frac{\partial f}{\partial p'}\right)^2 + \left(\frac{\partial f}{\partial q}\right)^2}} = \frac{-\eta}{\sqrt{1+\eta^2}} \tag{5a}$$

$$n_q^f = \frac{\frac{\partial f}{\partial q}}{\left\| \frac{\partial f}{\partial \underset{\sim}{\sigma}} \right\|} = \frac{\frac{\partial f}{\partial q}}{\sqrt{\left(\frac{\partial f}{\partial p'}\right)^2 + \left(\frac{\partial f}{\partial q}\right)^2}} = \frac{1}{\sqrt{1+\eta^2}} \tag{5b}$$

### 2.3 Dilatancy and Plastic Flow Rule

Based on theoretical analysis on irregular packing of uniform rods under horizontal and vertical stresses, Rowe (1962) proposed a stress-dilatancy relation as follows:

$$\sin\psi_m = \frac{\sin\varphi_m - \sin\varphi_{cs}}{1 - \sin\varphi_m \sin\varphi_{cs}} \tag{6}$$

where $\psi_m$ is the dilatation angle; $\varphi_m$ is the mobilized friction angle; and $\varphi_{cs}$ is the critical state friction angle.

Consider the difference in deformation mechanism and particle shape between uniform rods and actual granular soils, Xu and Song (2009) modified the stress-dilatancy equation as follows:

$$\sin\psi_m = \chi_d \frac{\sin\varphi_m - \sin\varphi_{cs}}{1 - \sin\varphi_m \sin\varphi_{cs}} \tag{7}$$

where $\chi_d$ is a dilation parameter.

Wan and Guo (1998) modified Rowe's stress-dilatancy equation by incorporating a state void-ratio index to describe the state-dependent behavior. The revised equation can be expressed as:

$$\sin\psi_m = \frac{\sin\varphi_m - (I_e)^{\vartheta} \sin\varphi_{cs}}{1 - (I_e)^{\vartheta} \sin\varphi_m \sin\varphi_{cs}} \tag{8}$$

where $\vartheta$ is a dilation parameter.

Based on Eqs. (7) and (8), a modified Rowe's stress-dilatancy equation is proposed, which can be expressed as:

$$\sin\psi_m = \chi_d \frac{\sin\varphi_m - (I_e)^{\vartheta} \sin\varphi_{cs}}{1 - (I_e)^{\vartheta} \sin\varphi_m \sin\varphi_{cs}} \tag{9}$$

The dilatancy $d$ is defined as:

$$d = -\frac{d\varepsilon_v^p}{d\varepsilon_s^p} \tag{10}$$

According to the study by Xiao et al. (2020), the dilatancy can be written as:

$$d = \frac{6 \sin \psi_m}{3 - \sin \psi_m} \tag{11}$$

Substituting Eq. (9) into Eq. (11) yields:

$$d = \frac{6\chi_d \sin \varphi_m - 6\chi_d \left( \frac{e}{e_\Gamma - \lambda_{cs}(p'_{cs}/p_a)^\xi} \right)^\vartheta \sin \varphi_{cs}}{3 - \chi_d \sin \varphi_m - (3 \sin \varphi_m - \chi_d) \sin \varphi_{cs} \left( \frac{e}{e_\Gamma - \lambda_{cs}(p'_{cs}/p_a)^\xi} \right)^\vartheta} \tag{12}$$

Hence, the plastic flow direction unit normal vector $\mathbf{n}^g = \left[n_p^g, n_q^g\right]^T$ can be expressed as:

$$n_{p'}^g = \frac{-d}{\sqrt{1+d^2}} = \frac{\frac{6\chi_d \sin \varphi_m - 6\chi_d (I_e)^\vartheta \sin \varphi_{cs}}{3 - \chi_d \sin \varphi_m - (3 \sin \varphi_m - \chi_d) \sin \varphi_{cs} (I_e)^\vartheta}}{\sqrt{1 + \left\{ \frac{6\chi_d \sin \varphi_m - 6\chi_d (I_e)^\vartheta \sin \varphi_{cs}}{3 - \chi_d \sin \varphi_m - (3 \sin \varphi_m - \chi_d) \sin \varphi_{cs} (I_e)^\vartheta} \right\}^2}} \tag{13a}$$

$$n_q^g = \frac{1}{\sqrt{1+d^2}} = \frac{1}{\sqrt{1 + \left\{ \frac{6\chi_d \sin \varphi_m - 6\chi_d (I_e)^\vartheta \sin \varphi_{cs}}{3 - \chi_d \sin \varphi_m - (3 \sin \varphi_m - \chi_d) \sin \varphi_{cs} (I_e)^\vartheta} \right\}^2}} \tag{13b}$$

### 2.4 Plastic Modulus

Based on (Li and Dafalias 2000; Xiao et al. 2020), the bounding stress ratio $M_b$ is defined as:

$$M_b = M_{cs}(I_e)^{-\beta} \tag{14}$$

where $\beta$ is a peak-failure parameter.

The critical state stress ratio $M_{cs}$ under the triaxial condition is defined as:

$$M_{cs} = \frac{6 \sin \varphi_{cs}}{3 - \sin \varphi_{cs}} \tag{15}$$

The plastic modulus $H_p$ can be expressed as a function of the elastic shear modulus ($G$), the stress ratio ($\eta$), the state void-ratio index ($I_e$), and the bounding stress ratio ($M_b$):

$$H_p = (h_1 - h_2 e_0)(I_\mathrm{e})^{-\beta} G \frac{(M_b - \eta)}{\eta} \frac{1}{(1+\eta)^2} \tag{16}$$

where $h_1$ and $h_2$ are the plastic modulus constants.

## 2.5 Constitutive Equations

Based on the elastic-plastic incremental theory, the strain-increment and stress-increment relations can be expressed as:

$$d\boldsymbol{\varepsilon} = d\boldsymbol{\varepsilon}^e + d\boldsymbol{\varepsilon}^p \tag{17a}$$

$$d\boldsymbol{\sigma} = \mathbf{D}^e\left(d\boldsymbol{\varepsilon} - d\boldsymbol{\varepsilon}^p\right) = \mathbf{D}^{ep}d\boldsymbol{\varepsilon} \tag{17b}$$

where $d\boldsymbol{\varepsilon}$, $d\boldsymbol{\varepsilon}^e$ and $d\boldsymbol{\varepsilon}^p$ are the total strain-increment, elastic strain-increment and plastic strain-increment, respectively; $d\boldsymbol{\sigma}$ is the stress-increment; $\mathbf{D}^e$ is the elastic stiffness matrix; and $\mathbf{D}^{ep}$ is the elastic-plastic stiffness matrix.

The elastic volumetric strain-increment $d\boldsymbol{\varepsilon}_v^e$ and elastic shear strain-increment $d\boldsymbol{\varepsilon}_s^e$ can be expressed as:

$$d\boldsymbol{\varepsilon}_v^e = \frac{dp'}{B_e} \tag{18a}$$

$$d\boldsymbol{\varepsilon}_s^e = \frac{dq}{3G_e} \tag{18b}$$

where $dp'$ is the mean effective stress increment; $dq$ is the deviatoric stress increment.

Hardin and Richart (1963) found that the elastic shear modulus $G_e$ correlates to the void ratio $e$ and mean effective stress $p'$ with the following equation:

$$G_e = G_0 p_a \frac{(2.97 - e)^2}{1 + e}\sqrt{\frac{p'}{p_a}} \tag{19}$$

where $G_0$ is a material constant.

The elastic bulk modulus $B_e$ is related to the elastic shear modulus $G_e$ and Poisson's ratio $\upsilon$, which can be written as:

$$B_e = G_e \frac{2(1+\upsilon)}{3(1-2\upsilon)} = G_0 \frac{(2.97 - e)^2}{1 + e}\frac{2(1+\upsilon)}{3(1-2\upsilon)}\sqrt{p' p_a} \tag{20}$$

The plastic strain-increment (i.e., $d\boldsymbol{\varepsilon}_v^p$ and $d\boldsymbol{\varepsilon}_s^p$) can be expressed as:

$$d\boldsymbol{\varepsilon}_v^p = d\lambda n_p^g \tag{21a}$$

$$d\boldsymbol{\varepsilon}_s^p = d\lambda n_q^g \tag{21b}$$

The plastic multiplier $d\lambda$ and elastic-plastic stiffness matrix $\mathbf{D}^{ep}$ can be expressed as:

$$d\lambda = \frac{\mathbf{n}_f^T : \mathbf{D}^e : d\boldsymbol{\varepsilon}}{H_p + \mathbf{n}_f^T : \mathbf{D}^e : \mathbf{n}_g} \tag{22a}$$

$$\mathbf{D}^{ep} = \mathbf{D}^{e} - \frac{\mathbf{D}^{e} : \mathbf{n}_g \otimes \mathbf{n}_f^T : \mathbf{D}^{e}}{H_p + \mathbf{n}_f^T : \mathbf{D}^{e} : \mathbf{n}_g} \tag{22b}$$

The proposed state-dependent model contains eleven parameters, including two elastic parameters $G_0$ and $\upsilon$; four critical state parameters $M_{cs}$, $\lambda_{cs}$, $\xi$, and $e_\Gamma$; two dilatancy parameters $\chi_d$ and $\vartheta$; one peak-failure parameter $\beta$; and two plastic modulus parameters $h_1$ and $h_2$. All model parameters can be determined based on triaxial data. Please referred to (Li and Dafalias 2000; Xiao et al. 2014b; Xiao and Liu 2017) for detailed calibration procedure.

## 3 Explicit Integration Algorithms

Numerical integration algorithm is key to implement a constitutive model in finite element codes. In this section, an explicit integration algorithm based on the modified Euler method with adaptive sub-stepping and error control, proposed by Sloan (2001), is applied to the sate-dependent constitutive model for rockfill materials. This algorithm automatically controls the size of the sub-increments by controlling the local errors of the stresses and internal variables throughout the integration process, which can improve the accuracy of the calculation while ensuring the efficiency.

In general, the adaptive sub-stepping explicit integration algorithm with error control mainly involves three parts: (i) For a given increment step and strain increment, a pseudo time $T$ $(0 \le T \le 1)$ is defined, and the total strain increment $d\boldsymbol{\varepsilon}$ is used to determine sub-increment $d\boldsymbol{\varepsilon}_n$ $(d\boldsymbol{\varepsilon}_n = \Delta T_n d\boldsymbol{\varepsilon})$ through the pseudo time sub-increment $\Delta T_n$; (ii) Through the first-order and second-order modified Euler solution, the local error of the stresses and internal variables can be determined by the difference between the two set of solutions. If the local error $R_n$ cannot less than the prescribed tolerance *STOL*, a smaller $\Delta T_n$ is needed and the size of each sub-step is automatically updated to a smaller one; (iii) If the local error $R_n$ is not great than *STOL*, the current sub-step could be accepted. Then, determine the step size for the next increment step, and update the stress and internal variables to begin the next incremental step calculation. The detail procedure of the adaptive sub-stepping explicit integration algorithm with error control is presented in Table 1.

## 4 Numerical Performance

To validate the proposed scheme, the state-dependent model with an adaptive sub-stepping explicit integration algorithm is implemented into a finite element code. The performance of the proposed model is evaluated at element level, and the element type used for this analysis is the eight-node solid element. The element is constrained at the bottom with lateral confining pressure. Axial strain is applied at the top at a constant rate to a final value of 20%. A series of drained triaxial compression tests for different rockfill materials are simulated with the state-dependent model to evaluate the performance of the numerical integration algorithm. All the model parameters are listed in Table 2.

**Table 1.** Procedure of the adaptive sub-stepping explicit integration algorithm with error control in the state dependent model

| Step | Description |
|---|---|
| 1 | Enter with initial stress $\boldsymbol{\sigma}_0$, strain increment $d\boldsymbol{\varepsilon}$, local error tolerance *STOL*, set initial pseudo time $T = 0$, initial pseudo time sub-increment $\Delta T_n = 1$, $d\boldsymbol{\varepsilon}_n = \Delta T_n d\boldsymbol{\varepsilon}$ |
| 2 | IF $T < 1$, then continue integration (go to step 3–11) |
| 3 | Calculate the first order ($j = 1$) trial stresses and the plastic strain increments<br>$d\boldsymbol{\sigma}_1 = \mathbf{D}^{ep}\left(\sigma_{n-1}, \varepsilon^p_{n-1}\right) : d\boldsymbol{\varepsilon}_n$<br>$d\boldsymbol{\varepsilon}^p_1 = d\lambda\left(\boldsymbol{\sigma}_{n-1}, \boldsymbol{\varepsilon}^p_{n-1}, d\boldsymbol{\varepsilon}_n\right) \cdot \mathbf{n}_g(\boldsymbol{\sigma}_{n-1})$ |
| 4 | Based on the first order trial stresses, update the stress and elastic-plastic stiffness matrix. Then calculate the second order ($j = 2$) trial stresses and the plastic strain increments<br>$d\boldsymbol{\sigma}_2 = \mathbf{D}^{ep}\left(\boldsymbol{\sigma}_{n-1} + d\boldsymbol{\sigma}_1, \boldsymbol{\varepsilon}^p_{n-1} + d\boldsymbol{\varepsilon}^p_1\right) : d\boldsymbol{\varepsilon}_n$<br>$d\boldsymbol{\varepsilon}^p_2 = d\lambda\left(\boldsymbol{\sigma}_{n-1} + d\boldsymbol{\sigma}_1, \boldsymbol{\varepsilon}^p_{n-1} + d\boldsymbol{\varepsilon}^p_1, d\boldsymbol{\varepsilon}_n\right) \cdot \mathbf{n}_g(\boldsymbol{\sigma}_{n-1} + d\boldsymbol{\sigma}_1)$ |
| 5 | Calculate new stresses and hardening parameters and hold them in temporary storage according to:<br>$\hat{\boldsymbol{\sigma}}_n = \boldsymbol{\sigma}_{n-1} + 0.5(d\boldsymbol{\sigma}_1 + d\boldsymbol{\sigma}_2)$<br>$\hat{\boldsymbol{\varepsilon}}^p_n = \hat{\boldsymbol{\varepsilon}}^p_{n-1} + 0.5\left(d\boldsymbol{\varepsilon}^p_1 + d\boldsymbol{\varepsilon}^p_2\right)$ |
| 6 | Determine the relative error $R_n$ for the current sub-step: $R_n = \max(R_\sigma, R_\varepsilon)$<br>$R_\sigma = \frac{1}{2}\sqrt{\frac{(d\boldsymbol{\sigma}_2 - d\boldsymbol{\sigma}_1):(d\boldsymbol{\sigma}_2 - d\boldsymbol{\sigma}_1)}{\hat{\boldsymbol{\sigma}}_n:\hat{\boldsymbol{\sigma}}_n}}$<br>$R_\varepsilon = \frac{1}{2}\sqrt{\frac{(d\boldsymbol{\varepsilon}^p_2 - d\boldsymbol{\varepsilon}^p_1):(d\boldsymbol{\varepsilon}^p_2 - d\boldsymbol{\varepsilon}^p_1)}{\hat{\boldsymbol{\varepsilon}}^p_n:\hat{\boldsymbol{\varepsilon}}^p_n}}$ |
| 7 | IF $R_n > STOL$, the current sub-step has failed and a smaller pseudo time step needs to be found by extrapolation. First compute<br>$\zeta = \max\left\{0.9\sqrt{STOL/R_n}, 0.1\right\}$<br>And then,<br>$\Delta T_n = \max\{\zeta\Delta T_n, \Delta T_{\min}\}, \ \Delta T_{\min} = 0.001$<br>Returning to step 2 |
| 8 | IF $R_n \le STOL$, the current sub-step is successful with regard to the local error control. Update the stresses and internal variables<br>$\boldsymbol{\sigma}_n = \hat{\boldsymbol{\sigma}}_n, \ \boldsymbol{\varepsilon}^p_n = \hat{\boldsymbol{\varepsilon}}^p_n$ |
| 9 | Extrapolate the size of the next sub-step<br>$\zeta = \max\left\{0.9\sqrt{STOL/R_n}, 0.1\right\}$<br>If the previous step failed, limit the step size growth further by enforcing<br>$\zeta = \min\{\zeta, 1.0\}$<br>Update pseudo-time and compute new step size<br>$T = T + \Delta T_{n+1}; \Delta T_{n+1} = \zeta\Delta T_n$ |

(*continued*)

**Table 1.** (*continued*)

| Step | Description |
|---|---|
| 10 | Ensure the next step size is not smaller than the minimum step size, and check the step size to avoid the pseudo time of integration does not higher than one<br>$\Delta T_{n+1} = \max\{\Delta T_{n+1}, \Delta T_{\min}\}$<br>$\Delta T_{n+1} = \min\{\Delta T_{n+1}, 1 - T\}$ |
| 11 | Exit with updated stress and internal variables with $T = 1$ |

**Table 2.** Model validation parameters

| Symbols | Latite ballast (Salim and Indraratna 2004) | Tacheng rockfill (Xiao et al. 2014a) | Basalt rockfill (Jia et al. 2017) |
|---|---|---|---|
| $G_0$ | 73 | 127 | 67.5 |
| $\upsilon$ | 0.25 | 0.25 | 0.25 |
| $M_{cs}$ | 2.04 | 1.66 | 1.79 |
| $\lambda_{cs}$ | 0.12 | 0.01 | 0.01 |
| $\xi$ | 0.6 | 0.7 | 0.7 |
| $e_\Gamma$ | 1.028 | 0.293 | 0.354 |
| $\chi_d$ | 0.75 | 1.12 | 1.15 |
| $\vartheta$ | 0.12 | 0.32 | 0.24 |
| $\beta$ | 3.96 | 0.92 | 1.02 |
| $h_1$ | 6.82 | 4.82 | 5.70 |
| $h_2$ | 8.35 | 12.25 | 14.52 |

Figure 2 presents the comparison between the triaxial compression data and the simulation results using the adaptive sub-stepping explicit integration algorithm for crushed latite ballast (Salim and Indraratna 2004). The initial void ratio of the specimens were $e_0 = 0.79, 0.76, 0.74$, and 0.72, corresponding to the initial effective confining pressure $p'_c = 50, 100, 200$, and 300 kPa, respectively. It is indicated that the proposed model is capable of predicting the stress-strain and volumetric behaviour with fair accuracy.

Figure 3 presents the comparision between the numerical simulations and experimental data for Tacheng rockfill (Xiao et al. 2014a) at different effective confining pressures. Drained triaxial tests under various pressures (from 400 to 1600 kPa) are simulated with the proposed constitutive model. The initial void ratio is 0.189 for all cases. The proposed model not only can describe the strain-hardening and volumetric contraction behaviors under confining pressures, but also can well capture the strain-softening and volumetric expansion behaviors under low confining pressures.

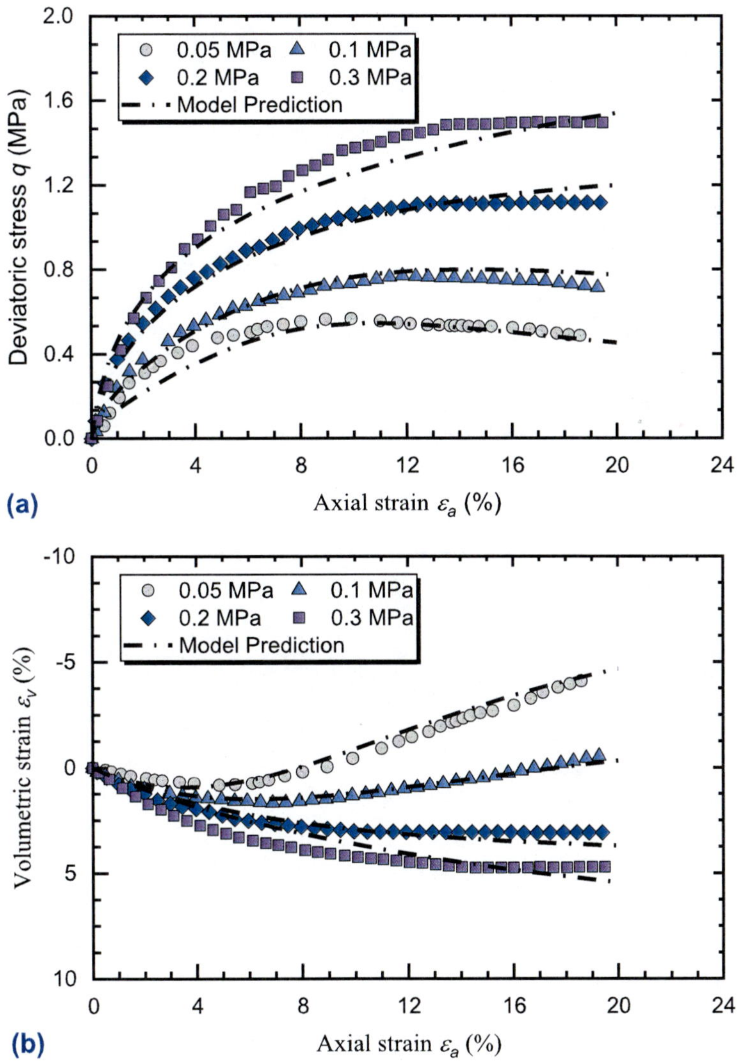


**Fig. 2.** Comparisons between numerical simulation and experimental data of Latite ballast: (a) deviatoric stress versus axial strain; (b) volumetric strain versus axial strain

Jia et al. (2017) performed a series of drained triaxial compression tests on basalt rockfill using a large-scale multifunctional triaxial apparatus. Experiments were conducted under various confining pressures, i.e. $p_c' = 500$ kPa, 1000 kPa, 1500 kPa, and 2000 kPa, while the initial void ratio were the same (0.262). Figure 4 indicates that all the model predications with the explicit algorithm are in good agreement with the experimental data.

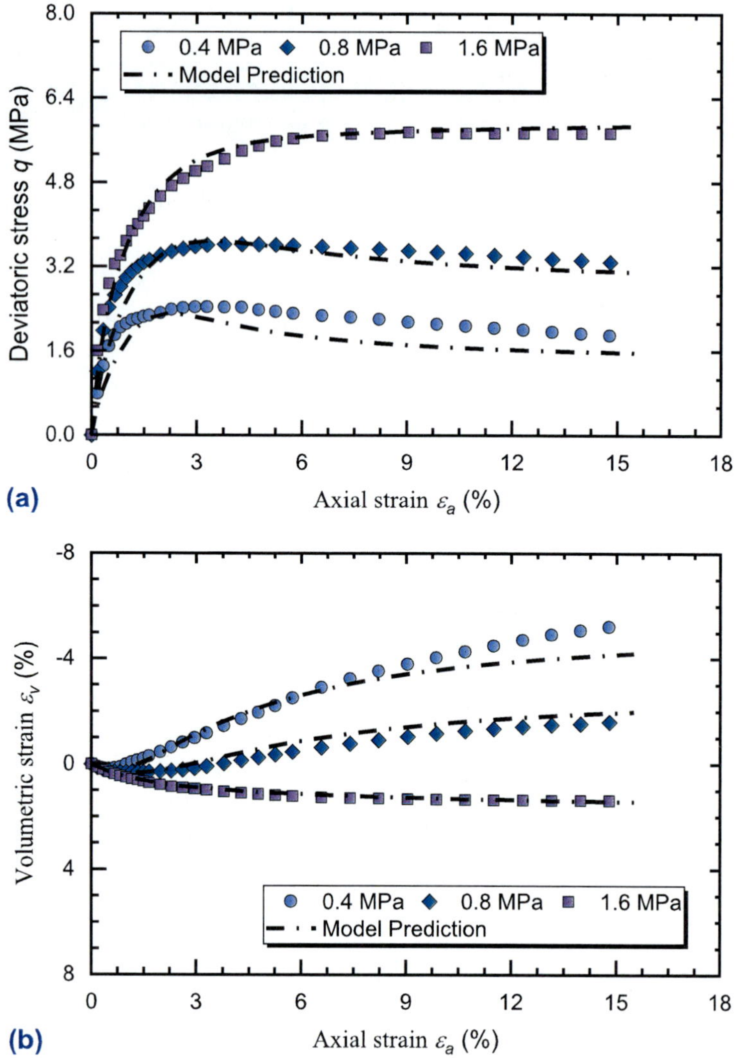


**Fig. 3.** Comparisons between numerical simulation and experimental data of Tacheng rockfill: (a) deviatoric stress versus axial strain; (b) volumetric strain versus axial strain

In explicit integration algorithm, the increment size and error tolerance (*STOL*) may influence the accuracy of the simulation results. For strain-softening and strain-hardening conditions, the effect of the increment size ($\Delta\varepsilon_a = 0.01, 0.001$, and $0.0001$) on the simulation results with different *STOLs* were analyzed. The model parameters are the same as Tacheng rockfill, as listed in Table 2. To assess the accuracy of the scheme in the state-dependent model, a global quantifiable error analysis is introduced to calculate

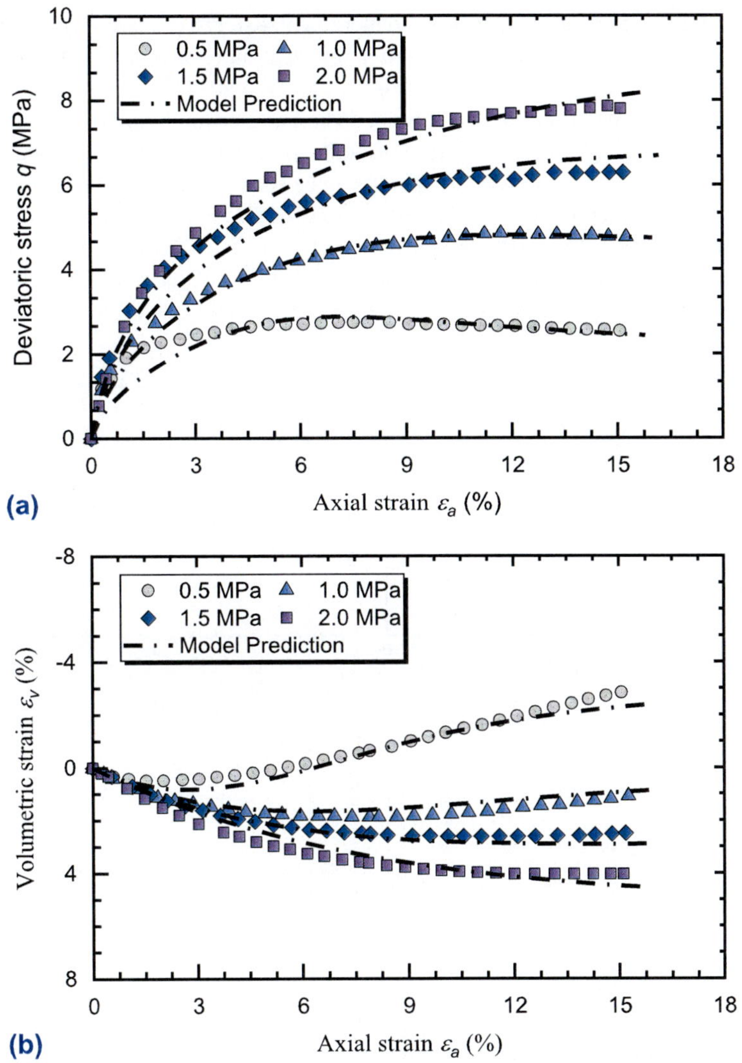


**Fig. 4.** Comparisons between numerical simulation and experimental data of Basalt rockfill: (a) deviatoric stress versus axial strain; (b) volumetric strain versus axial strain

the difference between a calculated stress and a reference stress. The relative error *RE* is defined as follows:

$$RE = \frac{\sum_{i=1}^{n} \left\| \sigma_i - \sigma_i^{ref} \right\|}{\sum_{i=1}^{n} \left\| \sigma_i^{ref} \right\|} \tag{23}$$

where $n$ is the number of stress points; $\sigma_i$ is the calculated stress of the $i$ th data point; $\sigma_i^{ref}$ is the reference stress of the $i$ th data point. In this study, the algorithm with 2000 total sub-increments and $STOL = 10^{-6}$ are used to obtain the reference results.

The simulation results are sensitive to incremental size. Compared with strain hardening conditions, the difference in simulation results is more obvious for strain softening when *STOL* is set to $10^{-1}$, as shown in Fig. 5(a, c). When $STOL = 10^{-5}$, the influence of the incremental size is almost negligible, and all of the simulation results are comparable, as shown in Fig. 5(b, d). Figure 6 presents the variations of the relative error RE for different incremental sizes and *STOLs*. It can be seen that the relative error decreases with a decrease in incremental size or a decrease in *STOL*. For the typical value of *STOL* ($10^{-3}$–$10^{-4}$), the relative error is approximately $10^{-3}$–$10^{-5}$ for different increment sizes. The results demonstrate the accuracy and stability of the adaptive sub-stepping explicit integration algorithm applied to the state-dependent constitutive model.

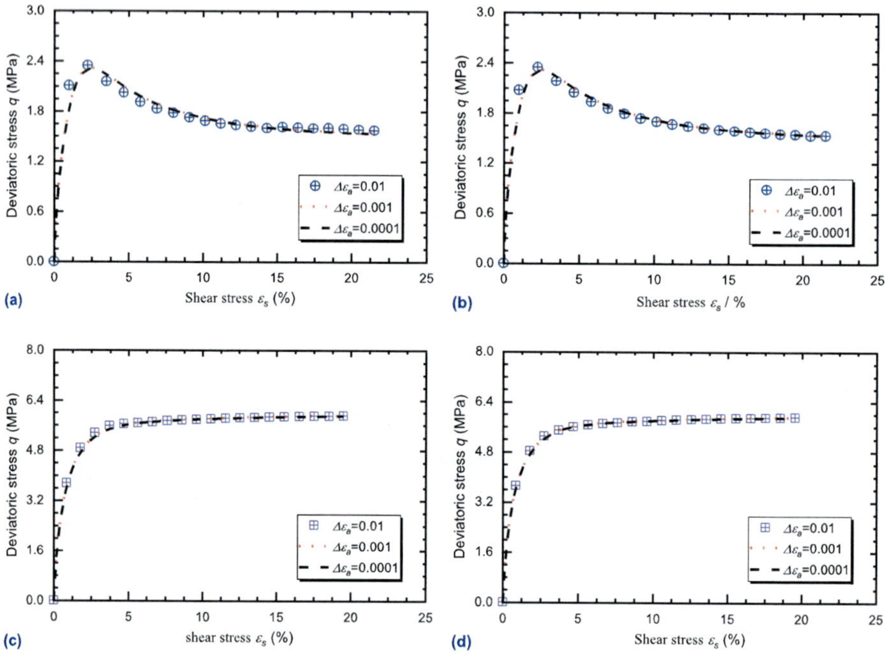


**Fig. 5.** Numerical simulation results for different increment sizes and *STOLs*: (a) $p'_c = 0.4$ MPa, $STOL = 10^{-1}$; (b) $p'_c = 0.4$ MPa, $STOL = 10^{-5}$; (c) $p'_c = 1.6$ MPa, $STOL = 10^{-1}$; (d) $p'_c = 1.6$ MPa, $STOL = 10^{-5}$

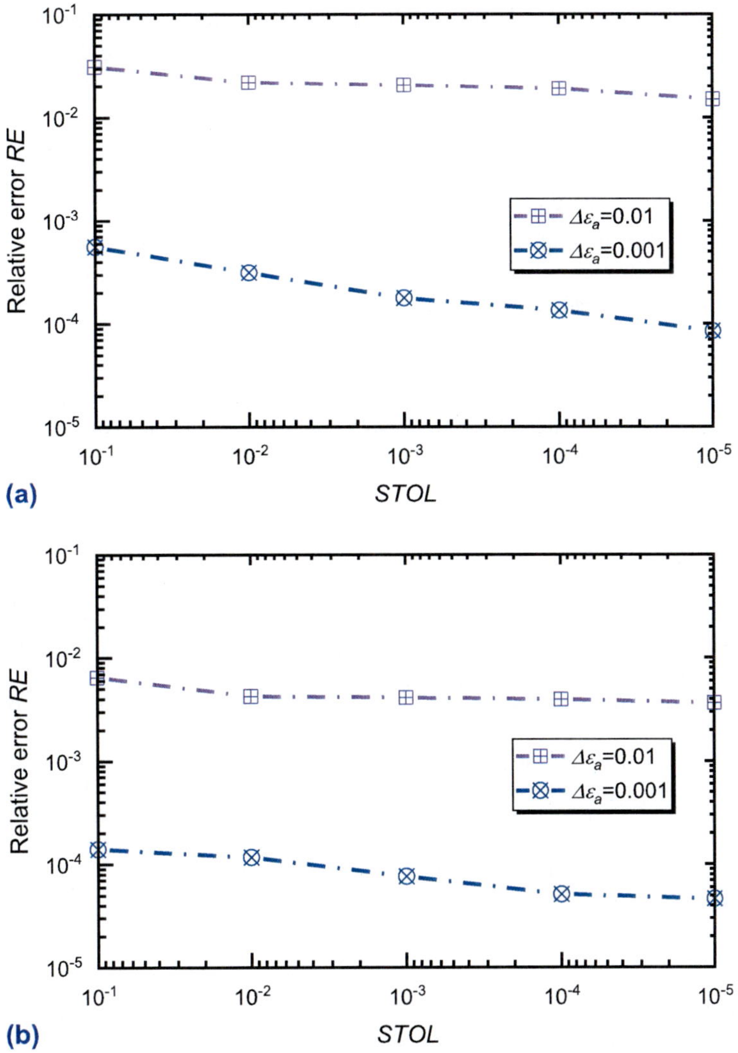


**Fig. 6.** Variations of the relative error for different increment sizes and *STOLs*: (a) $p_c' = 0.4$ MPa; (b) $p_c' = 1.6$ MPa

## 5 Conclusions

A state-dependent constitutive model was proposed for rockfill materials by incorporating a modified Rowe's stress-dilatancy equation. The state void-ratio index was incorporated into the Rowe's stress-dilatancy equation, bounding stress ratio, and plastic modulus. The model with an adaptive sub-stepping integration algorithm is implemented into a finite element code to predict behaviors of several typical rockfill materials. Major findings are summarized below:

(1) Comparisons between the element-level simulation results and experimental data demonstrate that the proposed model can well describe the strain hardening, strain softening, and dilatancy behaviors of rockfill materials in a wide range of densities and pressures. The feasibility and validity of the explicit integration algorithm on the state-dependent model for rockfill materials are verified.
(2) Through analyzing the simulation results for different incremental size and *STOLs*, the adaptive sub-stepping explicit integration algorithm for the state-dependent model performs well, in terms of both accuracy and convergence. The present work could provide a reference for the engineering application of rockfill materials constitutive models.

**Acknowledgments.** The authors would like to acknowledge the financial support from the National Science Foundation of China (Grant No. 51678094 and Grant No. 51922024), Sponsored by Natural Science Foundation of Chongqing, China (Grant No. cstc2019jcyjjqX0014).

## References

Been, K., Jefferies, M.G.: A state parameter for sands. Géotechnique **35**(2), 99–112 (1985)
Duncan, J., Chang, C.Y.: Nonlinear analysis of stress and strain in soils. J. Soil Mech. Found. Div. **96**, 1629–1653 (1970)
Hardin, B.O., Richart, F.E.: Elastic wave velocities in granular soils. J. Soil Mech. Found. **89**(1), 33–66 (1963)
Hu, C., Liu, H.: Implicit and explicit integration schemes in the anisotropic bounding surface plasticity model for cyclic behaviours of saturated clay. Comput. Geotech. **55**, 27–41 (2014)
Jia, Y.F., Xu, B., Chi, S.C., Xiang, B., Zhou, Y.: Research on the particle breakage of rockfill materials during triaxial tests. Int. J. Geomech. **17**(10), 04017085 (2017)
Li, X.S., Dafalias, Y.F.: Dilatancy for cohesionless soils. Géotechnique **50**(4), 449–460 (2000)
Rowe, P.W.: The stress-dilatancy relation for static equilibrium of an assembly of particles in contact. Proc. Roy. Soc. Lond. A **269**(1339), 500–527 (1962)
Salim, W., Indraratna, B.: A new elastoplastic constitutive model for coarse granular aggregates incorporating particle breakage. Can. Geotech. J. **41**(4), 657–671 (2004)
Sheng, D.C., Sloan, S.W., Gens, A., Smith, D.W.: Finite element formulation and algorithms for unsaturated soils. Part I: theory. Int. J. Numer. Anal. Meth. Geomech. **27**(9), 745–765 (2003)
Sloan, S.W.: Refined explicit integration of elastoplastic models with automatic error control. Eng. Comput. **18**(1/2), 121–194 (2001)
Varadarajan, A., Sharma, K.G., Abbas, S.M., Dhawan, A.K.: Constitutive model for rockfill materials and determination of material constants. Int. J. Geomech. **6**(4), 226–237 (2006)
Wan, R.G., Guo, P.J.: A simple constitutive model for granular soils: modified stress-dilatancy approach. Comput. Geotech. **22**(2), 109–133 (1998)
Wang, Z.L., Dafalias Yannis, F., Li, X.S., Makdisi Faiz, I.: State pressure index for modeling sand behavior. J. Geotech. Geoenviron. Eng. **128**(6), 511–519 (2002)
Xiao, Y., Liu, H.: Elastoplastic constitutive model for rockfill materials considering particle breakage. Int. J. Geomech. **17**(1), 04016041 (2017)
Xiao, Y., Liu, H., Chen, Y., Jiang, J.: Bounding surface model for rockfill materials dependent on density and pressure under triaxial stress conditions. J. Eng. Mech. **140**(4), 04014002 (2014a)
Xiao, Y., Liu, H., Chen, Y., Jiang, J., Zhang, W.: Testing and modeling of the state-dependent behaviors of rockfill material. Comput. Geotech. **61**(3), 153–165 (2014b)

Xiao, Y., Sun, Z.C., Stuedlein, A.W., Wang, C.G., Wu, Z.J., Zhang, Z.C.: Bounding surface plasticity model for stress-strain and grain-crushing behaviors of rockfill materials. Geosci. Front. **11**(2), 495–510 (2020)

Xu, M., Song, E.: Numerical simulation of the shear behavior of rockfills. Comput. Geotech. **36**(8), 1259–1264 (2009)

Yao, Y., Liu, L., Luo, T.: A constitutive model for granular soils. Sci. China Technol. Sci. **61**(10), 1546–1555 (2018). https://doi.org/10.1007/s11431-017-9205-8

Yin, Z.Y., Hicher, P.Y., Dano, C., Jin, Y.F.: Modeling mechanical behavior of very coarse granular materials. J. Eng. Mech. **143**(1), C4016006 (2017)

Zhou, A., Zhang, Y.: Explicit integration scheme for a non-isothermal elastoplastic model with convex and nonconvex subloading surfaces. Comput. Mech. **55**(5), 943–961 (2015). https://doi.org/10.1007/s00466-015-1144-3

# Expander Bodies Provide Increased Axial Resistance to Full Drilled Displacement Piles for World Trade Center Towers in Bolivia

Antonio Marinucci[1(✉)], Mario Terceros Arce[2], and Mario A. Terceros Herrera[2]

[1] V2C Strategists, LLC, New York, NY, USA
amarinucci@v2cstrategists.com
[2] Incotec S.A., Santa Cruz de la Sierra, Bolivia
{mta,math}@incotec.cc

**Abstract.** To support column footings, slabs, elevator shafts, and shear walls for the World Trade Center building in Santa Cruz, Bolivia, about 350 full-displacement bored piles (FDPs) were designed and installed for an unfactored load of 255 tonne (2,500 kN). The piles were outfitted with an Expander Body (EB) to improve the axial bearing. The borehole diameter of each pile was about 500 mm and the EB was sized to be able to expand to maximum diameter of about 800 mm. A static head-down axial compression loading test was performed to evaluate the axial resistance and load-displacement response of the FDP+EB system. The test results showed that equipping an FDP pile with an EB significantly improved the bearing and enabled the FDP+EB pile to be considerably shortened—from 39 m to 29 m. The shortening of the piles saved the project about 25% in pile construction costs and about 40% in construction time. This paper provides an overview of the EB system and design methodology, summarizes the construction monitoring, and presents the results of the static loading test.

**Keywords:** Expander Body · Bored pile · Pressuremeter · Pile loading test · Quality control · Axial resistance · Post-grouting

## 1 Introduction

An Expander Body (EB) is an inflatable steel-shell unit added to the toe of a cast-in-place or vibrated pile or ground anchor and is inflated via grout injection after the installation of the pile/anchor to increase its axial resistance. As reported by various practitioners (e.g., Sellgren et al. 1989; Terceros et al. 1995; Fellenius and Terceros 2014), the EB system has been used successfully in a wide range of applications, including for support of high-rise buildings, bridge structures, industrial facilities, retaining walls, and underpinning works (Fig. 1). The following discussion describes the principles and process for an axially compression loaded pile+EB system. Similar principles apply for ground anchors axially loaded in tension.


H. Khabbaz et al. (Eds.): GeoChina 2021, SUCI, pp. 94–109, 2021.
https://doi.org/10.1007/978-3-030-79650-1_8

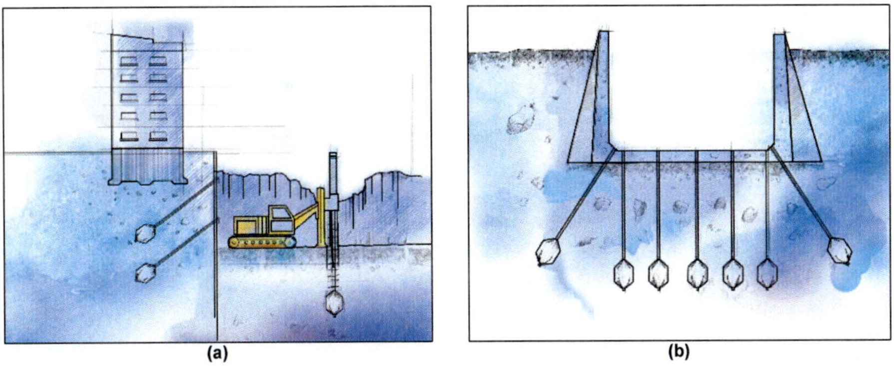


**Fig. 1.** Representative applications: (a) axial support with piling and lateral support with ground anchors, and (b) uplift resistance with ground anchors

## 2 Basic Principles

The EB consists of three integrated components: a shell (a folded steel sheet) with a circular cross section, a central steel reinforcement section, and grouting arrangements. After the borehole has been formed using the applicable construction piling technique (e.g., FDP), an EB assembly (e.g., reinforcement steel/sections, EB attachment, post-grouting tube, and instrumentation) is inserted into the borehole before or after the placement of grout/concrete (Fig. 2). The EB is expanded by injecting grout and thereafter grout is injected also below the EB to ensure a stiff base. The timing of the grouting for the EB is independent of the compressive strength of the grout/concrete because the thin layer of grout/concrete surrounding the EB, from the grouting of the borehole, will not interfere with the grouting of the EB.

**Fig. 2.** Schematic illustrating installation and expansion processes for piling applications

The EB shell expands into the surrounding soil as grout is injected under pressure (i.e., primary grouting). During the initial phases of grouting, the disturbed soil around the device is recompressed first and then, as grouting is continued, the soil is compressed. The expansion densifies the soil and results in increased shear resistance at the soil-EB interface and an increase in the stiffness of the soil radially away from the EB to a distance of 3 to 4 diameters of the EB. The shell undergoes two distinct responses during grout injection: (1) expansion outward into the surrounding soils and (2) a reduction in the effective length (i.e., shortening) of the EB. The diameter of the EB prior to grout injection is about 120 mm and it can expand to 400 to 800 mm in diameter depending on the model type and geotechnical conditions of the in-situ soils. The incorporation of post-grouting below the pile toe (i.e., secondary grouting), as shown in the rightmost schematic in Fig. 2, is intended to enhance performance and addresses potential soil decompression due to the shortening of the element. Monitoring of the grouting and post-grouting process serves as a quality control measure, similar to the pressuremeter test.

## 3 Design Methodology

The design for axial compression loading is essentially separated into two components: (1) estimation of the nominal side resistance provided by the pile (upper portion) and (2) estimation of the nominal side and toe resistances provided by the EB (lower portion). The estimation of the nominal side resistance provided by the pile depends on the pile type and specific installation method. The details and procedures for designing drilled shafts/bored piles, continuous flight auger piles, drilled displacement piles, and micropiles piles can be found in the technical literature, such as Brown et al. (2018), Brown et al. (2007), Fellenius (2018), NeSmith (2002), and Sabatini et al. (2005). For the purposes of this paper, the methodology presented below focuses on the design considerations pertaining to the EB portion of the piling element only. A more thorough discussion of the design process is provided in Marinucci et al. (2020); however, the pertinent design details are summarily presented below.

The nominal axial compression resistance of an EB Body is estimated using Allowable Stress Design (ASD), as shown in Eq. (1). As discussed in Marinucci et al. (2020), the design of an EB was based on cavity expansion theory and pressuremeter testing; therefore, the engineering parameters obtained during the site characterization program (e.g., SPT $N_{60}$-values, CPT $q_t$ and $f_s$, and PMT $p_{lim,PMT}$) must be converted to equivalent pressuremeter values for use in the design equations. For the design phase, the nominal side and toe resistances ($Q_{s,EB,d}$ and $Q_{t,EB,d}$) are calculated from Eqs. (2) and (3), respectively, using the EB surface area and toe area (Table 1) for a fully expanded model ($A_{sfe,EB}$ and $A_{tfe,EB}$, respectively). The post-expansion nominal shaft and toe resistances ($Q_{s,EB,f}$ and $Q_{e,EBI,f}$) are computed using Eqs. (4) and (5), respectively, which are determined using the limit pressure ($p_{lim,EB}$) measured during the grouting operation. Based on the volume of grout injected, the effective diameter of the EB (the effective EB surface area and toe area: $A_{s,EB,f}$ and $A_{t,EB,f}$, respectively), which can be determined using the appropriate calibration curve for the particular model used (Fig. 3).

$$Q_{ta,EB} = \frac{Q_{s,EB,i}}{FS_s} + \frac{Q_{t,EB,i}}{FS_t} \quad (1)$$

$$Q_{s,EB,d} = 2p_{lim,PMT}A_{sfe,EB}tan\delta \quad (2)$$

$$Q_{t,EB,d} = \{k[2p_{lim,PMT} - \sigma_h] + \sigma_v\}A_{tfe,EB} \quad (3)$$

$$Q_{s,EB,f} = p_{lim,EB}A_{s,EB,f}\,tan\delta \quad (4)$$

$$Q_{t,EB,f} = kp_{lim,EB}A_{t,EB,f} \quad (5)$$

where:
$Q_{ta,EB}$ = Total allowable axial resistance provided by the EB
$Q_{s,EB}$ = Nominal, unfactored axial side resistance provided by the EB
$Q_{t,EB}$ = Nominal, unfactored axial toe resistance provided by the EB
$i$ = Condition being analyzed: design ($i$) or as constructed ($f$)
$FS_s$ = Factor of Safety for side resistance
$FS_t$ = Factor of Safety for toe resistance
$Q_{s,EB,d}$ = Design nominal, unfactored axial side resistance (fully expanded EB)
$Q_{t,EB,d}$ = Design nominal, unfactored axial toe resistance (fully expanded EB)
$p_{lim,d}$ = Design limit pressure based on in-situ measurements = $2 * p_{lim,PMT}$
$p_{lim,PMT}$ = Limit pressure determined using the PMT
$A_{sfe,EB}$ = Side surface area of the (assumed) fully expanded EB
$\delta$ = Friction angle between dissimilar materials: soil and steel = 10 degrees
$k$ = Empirical bearing capacity factor (Clarke, 1995 or derived based on load testing)
$\sigma_h$ = Total horizontal stress acting at the base of the EB
$\sigma_v$ = Total vertical stress acting at the base of the EB
$A_{tfe,EB}$ = Area of the toe of the (assumed) fully expanded EB
$Q_{s,EB,f}$ = Nominal, unfactored axial side resistance of as-constructed EB
$p_{lim,EB}$ = Limit pressure from measurements from grouting operation (real limit pressure)
$A_{s,EB,f}$ = Effective side surface area based on volume of grout injected and calibration curve
$Q_{t,EB,f}$ = Nominal, unfactored axial toe resistance of as-constructed EB
$A_{t,EB,f}$ = Effective area of the toe based on volume of grout injected and calibration curve

**Table 1.** Details of select EBI models

| EBI type | Pre-expansion | | Post-expansion (fully expanded) | | | | |
|---|---|---|---|---|---|---|---|
| | Diameter $D_{it,EB}$ (mm) | Length $L_{i,EB}$ (m) | Diameter $D_{ft,EB}$ (mm) | Length $L_{f,EB}$ (m) | Area at toe $A_{tfe,EB}$ (sq m) | Side surface area $A_{sfe,EB}$ (sq m) | Volume $V_{fe,EB}$ (liter) |
| EBI412 | 120 | 1.2 | 400 | 0.92 | 0.13 | 1.16 | 110 |
| EBI615 | 125 | 1.5 | 600 | 1.10 | 0.28 | 2.10 | 250 |
| EBI815 | 145 | 1.5 | 800 | 1.05 | 0.50 | 2.64 | 480 |

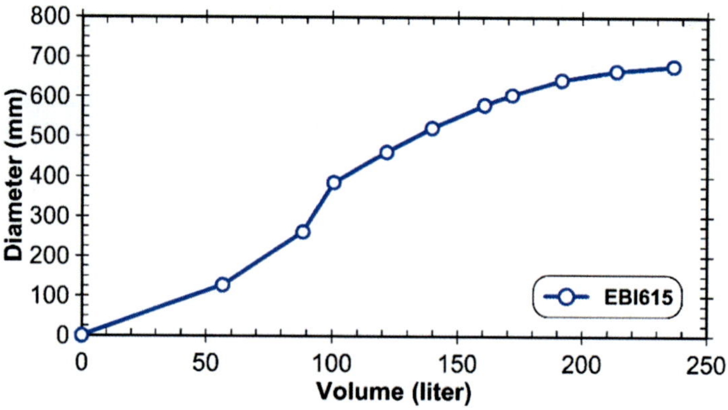


**Fig. 3.** Example of a calibration curve (EBI615 model shown)

## 4 Soil Response During Expansion

The pressure required to begin the expansion of the EB outward is equal to the horizontal stress acting on the EB from the surrounding soil. During the early stages of grouting, the disturbed soil around the device is recompressed (Zone *I* in Fig. 4). The relative size of Zone *I* (i.e., until $v_0$, $p_0$ is reached) is qualitatively related to soil type and density as well as to the amount of disturbance caused from the installation process. As additional grout is injected, the device continues to expand and compresses the in-situ soil (Zone *II*, the pseudo-elastic zone). As shown in Fig. 2, the amount of compression or densification that occurs in zone *II* (i.e., between $v_0$, $p_0$ and $v_f$, $p_f$) is a function of the installation method used (i.e., from a non-displacement pile [NDP] to a full displacement pile [FDP]).

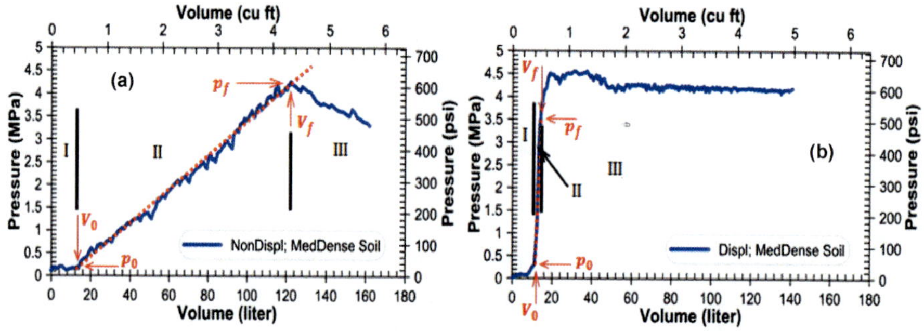


**Fig. 4.** Representative pressure-volume curves in medium dense sandy soil for an (a) NDP+EB and (b) FDP+EB (after Marinucci et al. 2020).

By evaluating the behavior trends of the pressure during injection, the EB system can detect the alteration to the stress state of the soil caused by the installation method. For an EB installed in a non-displacement pile (NDP+EBI), where the soil was likely stress relieved or decompressed during the construction process, the expansion of the EB requires a greater grout volume (i.e., the soil requires more compaction) to achieve its peak pressure. When an EB is installed in a full-displacement pile (FDP+EBI), it has been observed that the peak injection pressure is achieved shortly after beginning the expansion process, which is an effect of the mechanical improvement to the soil that had already occurred during the installation of the pile. With continued injection and expansion, the yield pressure ($p_f$; $p_{lim,EB}$ in Eqs. 4 and 5) is achieved (at the interface between Zones *II* and *III*), where large deformations and shearing of the soil are realized with continued grouting (Zone *III*, the plastic zone).

## 5 Case History

The groundbreaking ceremony for the new World Trade Center in Santa Cruz de la Sierra, Bolivia (WTC) took place in 2017, and the construction of the twin structures is currently ongoing. The development comprises two 30-story towers connected by four enclosed bridge structures. When completed, there will be more than 90,000 sq.m of above ground floor space along with four subterranean levels that will house more than 750 parking spaces. The WTC will provide private office space on 26 stories along with a conference and exhibition center, restaurants, auditorium, and other meeting spaces. The overall footprint of the development is approximately 128 m by 57 m. During the next phase of the development, a hotel and a convention center are expected to be constructed adjacent to the WTC structures.

### 5.1 Subsurface Conditions

The site characterization program was performed in two phases. The first investigation consisted of performing 9 soil borings, each to a depth of about 25 m, with standard penetration testing (SPT) blow counts (N-indices) performed at intervals of about 1 m (SiteCal 2015). The second phase of the site investigation consisted of 6 additional soil borings (with SPT blow counts) to depths between 35 and 40 m, 4 cone penetration test soundings with pore water pressure measurement (CPTu) to a depth of about 10 m, and 2 multichannel analyses of surface waves (MASW). (Only the records of the soil borings and SPT measurements were available for inclusion in this paper.) At the time the soil borings were performed, the groundwater table was found to be at a depth of about 14.5 m below the ground surface. An aerial view of the approximate locations of the project site and of the different in-situ testing performed are shown in Fig. 5. The profiles of SPT N-indices are presented in Fig. 6 and Fig. 7.

The highly heterogeneous soil deposit beneath the towers consists of alternating layers of sands (SP, SP-SM, SM, and SC), silts (ML), and clays (CL) to a depth of about 35 m. Loam to sandy fill (with organics) was observed in the uppermost layer and varied in thickness from about 1 to 4 m. Underlying the near surface layer to a depth of about 15 m to 17 m was a layer of mainly silty sands (SP-SM and SM) with interbedded layers of silt or clay. The sands were underlain by a layer of lean clay (CL) down to a depth of about 35 m, with thin lenses of sands and silt. The lean clay was more prominent on the southern side at near surface depths as well as at depths in excess of 15 m; however, the lean clay appeared to taper off toward the northern side of the project. Based on the details provided in the boring logs, there was a layer of varying thickness in which refusal by the sampler was noted. The layer of refusal was found to be shallower and slightly thinner toward the southern side of the project site (B-1 to B-3), deeper and thicker toward the northern soil borings (B-7 to B-9). The soil properties used for the design of the foundations are shown in Table 2.

## 5.2 Foundation Design and Installation

Based on the results of a technical-cost benefit analysis, the contractor offered an alternative technical concept and redesigned the deep foundations. The proposed foundation scheme employed full displacement piles (FDPs): FDPs without EBs to support the lighter loaded areas and FDP+EB with EBs attached to provide support in the more heavily loaded areas. The overarching concept for the alternative was to optimize the foundation system while providing a practical solution that ensured more uniform and acceptable settlements across the structures and towers. The FDPs were designed to be 360 mm in diameter with an assigned working load of 60 tonne (590 kN) and embedded into the medium dense to dense SM layer at or below El. +403.0 m, for a minimum length of about 9 m. The FDP+EBI were designed to be 500 mm in diameter with an assigned working load of 255 tonne (2500 kN) and embedded into the dense SP layer at or below El. +380.0 m, for a minimum length of about 35 m. The contractor also recommended performing one head-down static loading test to assist in evaluating the load-displacement performance of the piles.

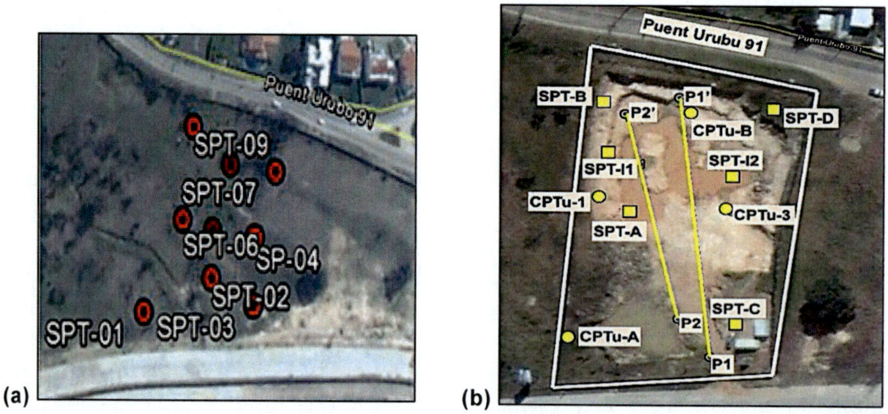


**Fig. 5.** Aerial view of project location and site investigation locations, as shown in (a) SiteCal (2015) geotechnical report and (b) Incotec (2017) report

The method used to design the FDPs was performed in general accordance with the discussion and recommendations by NeSmith (2002) and Brettmann and NeSmith (2005). However, the methods were adjusted slightly to incorporate local soil conditions and strength parameters determined from an analysis of loading tests performed on different projects in the region. The design of the EB section of the piles was performed as described above in Sect. 3. In general, the ultimate unit toe ($q_t$) and side ($q_s$)

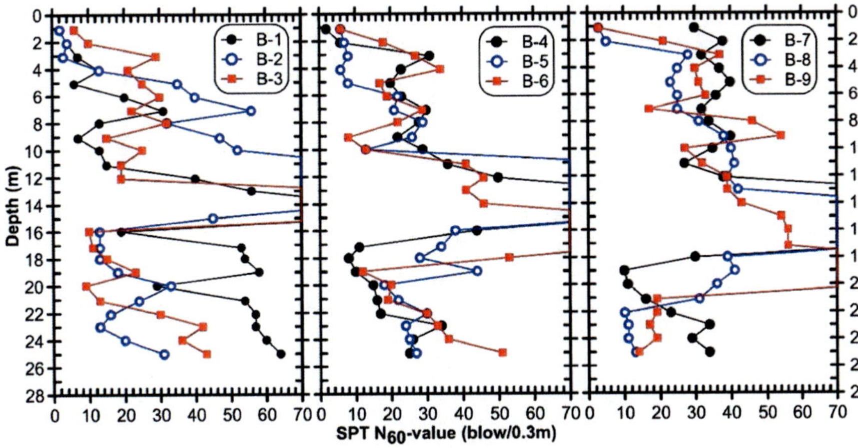


**Fig. 6.** Profiles of SPT $N_{60}$-values: soil borings B1 to B9 (after SiteCal 2015)

**Table 2.** Soil properties used for the design of the FDPs and FDP+EBs

| Elevation (m) | Soil type | Unit weight $\gamma_t$(kN/m$^3$) | Cohesion $c$ (kPa) | Friction angle $\phi$ (deg) |
|---|---|---|---|---|
| El. +411.5 to El. +405.0 | Med. dense SM | 18.5 | – | 32 |
| El. +405.0 to El. +403.0 | Inorganic CL | 17.7 | 20 | 15 |
| El. +403.0 to El. +397.5 | Med. dense to dense SM | 21.9 | – | 35 |
| El. +397.5 to El. +381.0 | CL with SP-SM | 17.7 | 26 | 20 |
| El. +381.0 to El. +375.0 | Dense SP | 22.7 | – | 38 |

Note: Elevation of ground surface (final, post-construction) = El. +418.50 m
Elevation of top of finished slab (lowest level) = El. +411.50 m

resistances of the FDPs were computed using Eqs. (6) and (7), respectively. As the soil along the length of the pile above the EBs were mainly fine-grained and transitional soils (i.e., silt and clay), the constants $w_e$ and $w_s$ were equal to zero. After the preliminary design of the piles and spacing at each location was completed, the commercial software program, *UniPile5* (Goudreault and Fellenius 2014), was used to evaluate the load-displacement behavior of the foundation system. Once the design was finalized, the maximum anticipated settlements of the foundation system were computed for three different construction stage/cases. Long-term settlement was estimated using the commercial software program, *UniSettle4* (Goudreault and Fellenius 2011), and showed that about 37 mm of displacement would occur at the center of each tower structure and that the differential settlements were smaller than L/500 (the permissible ratio). Details of the two pile types are provided in Fig. 8, while the layout of the foundation piles and structural pile caps/footings are shown in Fig. 9. In total, about 430 piles were installed of which about 350 were FDP+EBs.

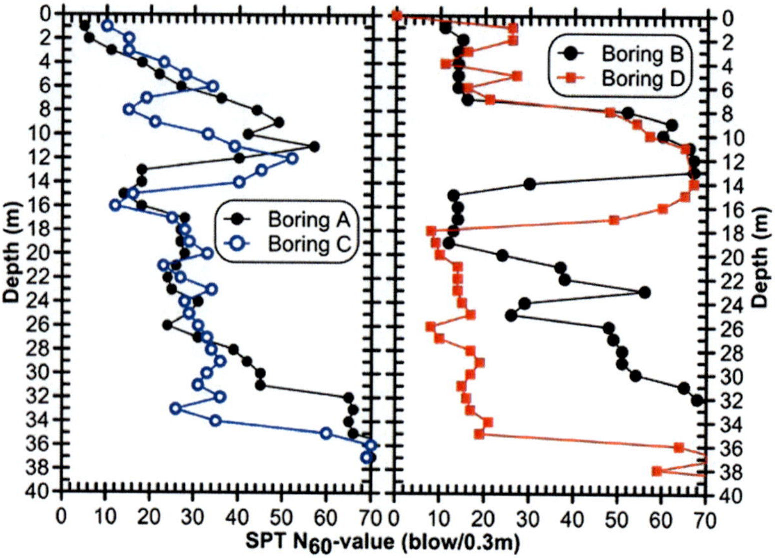


**Fig. 7.** Profiles of SPT $N_{60}$-values: soil borings A to D (after Incotec 2017)

$$q_e = 0.19N_m + w_e + w_{EB} \quad (N < 50\,\text{blow}/0.3\,\text{m}) \tag{6}$$

$$q_{si} = 0.005N_i + w_{si} \quad (N < 50\,\text{blow}/0.3\,\text{m}) \tag{7}$$

where:

$q_e$ = Ultimate unit toe resistance (MPa)
$N$ = Uncorrected blow count (N-index) as recorded in the field
$N_{60}$ = SPT blow count (N-index) corrected to 60% efficiency
$N_m = 0.25N_{60,0} + 0.25N_{60,1} + 0.5N_{60,2}$
(representative $N_{60}$-indices within a distance of 4 pile diameters below the base of the pile, where $N_{60,0} \leq 50$ (Fleming and Thorburn, 1983))
$N_{60,0}$ = Average $N_{60}$ from the pile toe to a depth 4 toe diameters below the toe
$N_{60,1}$ = Minimum $N_{60}$ from the pile toe to a depth 4 toe diameters below the toe
$N_{60,2}$ = Average $N_{60}$ from base of the pile to a depth of 4 toe diameters above the toe
$w_e$ = Constant; depends on the gradation and angularity of the soil
= 0 for soils with uniform content of rounded particles or for fine-grained soils
= 1.34 MPa for well-graded soils and angular particles with less than 10% fines
$w_{EB}$ = Constant = 0.95 MPa (Incotec, 2020)
$q_s$ = Ultimate unit side resistance (MPa)
$i$ = Layer for which unit side resistance is being computed
$w_s$ = Constant; depends on the gradation and angularity of the soil
= 0 for soils with uniform content of rounded particles or for fine-grained soils
= 0.05 MPa for well-graded soils and angular particles with less than 10% fines

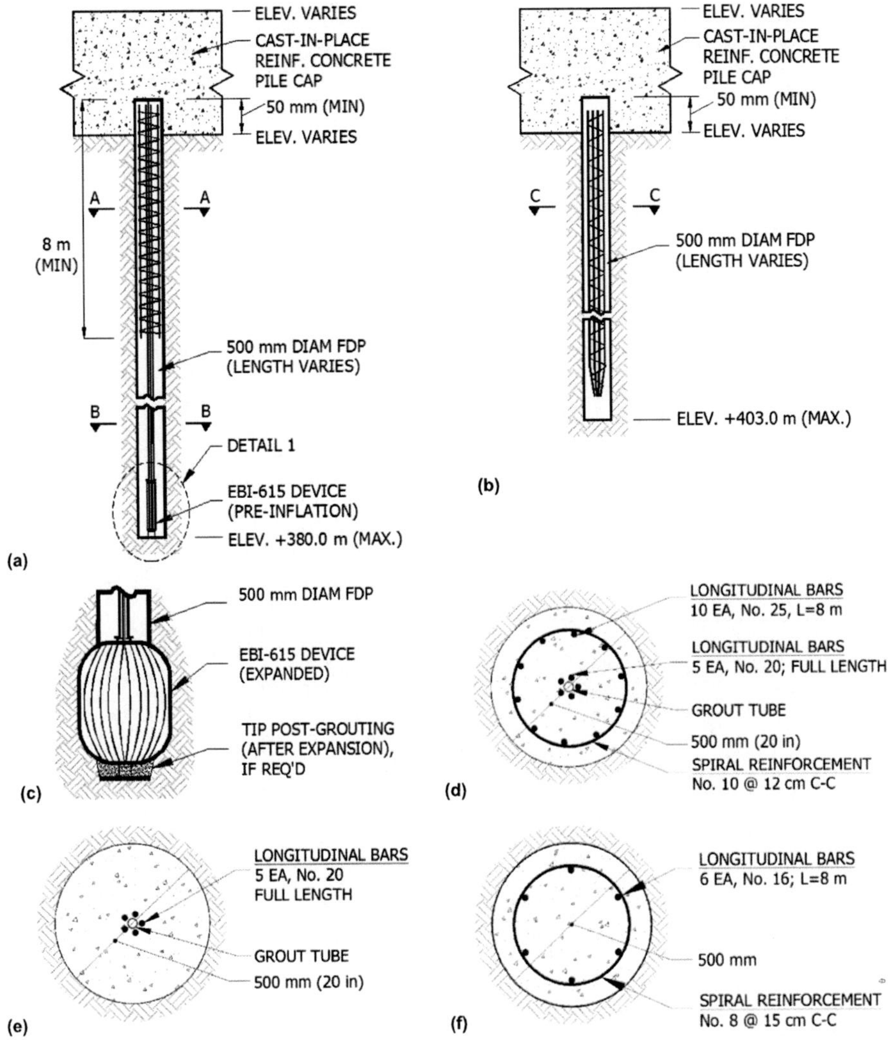


**Fig. 8.** Details of the foundation piles: (a) elevation view of an FDP+EB (pre-inflation), (b) elevation view of an FDP, (c) detail 1 (post-expansion of EB), (d) section A-A, (e) section B-B, and (f) section C-C

The foundation piles were installed using a dedicated Bauer drill rig outfitted with customized FDP drill tooling. For each pile, the drill string was first advanced to the desired depth, and then, the internal steel reinforcement and EB (where applicable) were inserted into the hollow stem portion of the drill tool. Then, concrete was pumped into the top and out of the bottom of the hollow stem, as the drill tool was extracted (the sacrificial drill bit remained in place at the bottom of the pile). Throughout the drilling and concreting processes, the drill string provided stability to the borehole until the concrete was placed. Once the drill tool was fully removed from the hole, the outer steel

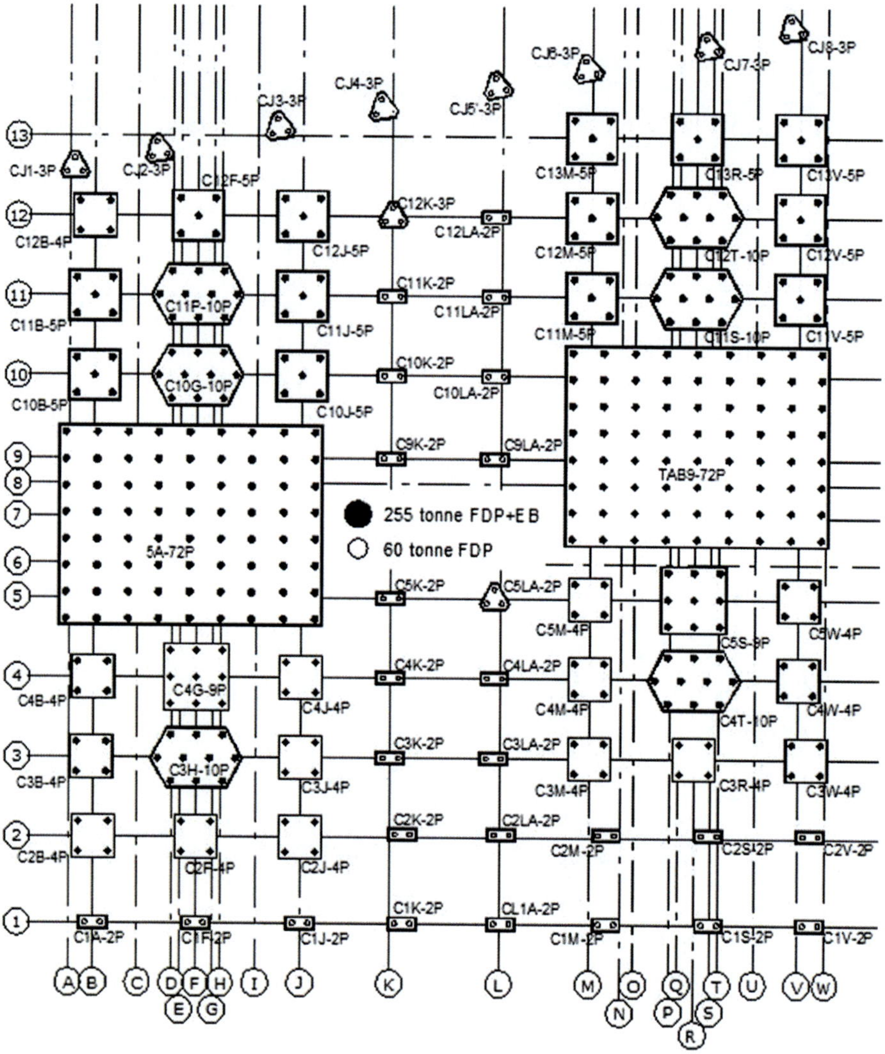


**Fig. 9.** Plan view of the contractor-designed foundation alternative

reinforcement section with the EB was inserted into the fluid concrete. The EBs were injected with grout at least one day after the pile was installed.

## 5.3 Static Loading Test – Production Pile

The construction included a special production test pile, C11J-P03, installed to a depth of 35 m. The as-constructed geometry of the test pile and the estimated axial resistances are shown in Table 3. A graph delineating the relationship between the measured injection pressure and injected grout volume for the test pile is shown in Fig. 10. Following expansion, the as-constructed axial resistance of the EB was estimated using the limit pressure, $p_{lim,EB}$, that was measured during the injection process. The toe and side surface areas ($A_{t,EB,f}$ and $A_{s,EB,f}$) used to estimate the axial resistance after expansion are defined by the measured injected grout volume along with the appropriate calibration curve for the EB model 815. As shown in Table 3, the equivalent diameter of the EB after inflation was about 800 mm, which is about 60% greater than the installed diameter of 500 mm of the overall pile. The side resistance of the EB alone was estimated to be about 550 kN acting over a length of about 1.05 m, compared with a side resistance from the FDP (above the EB) of about 1,300 kN acting over a length of about 28 m. The axial side resistance over a length of about 1.0 m above the EB was greatly improved due to the densification effect of the EB expansion. Moreover, the toe resistance was estimated to be about 710 kN due to the increased toe diameter.

**Table 3.** As-constructed geometry of the FDP+EB test pile

| Test pile | Pile length (m) | Injection – EB (max.) | | | Estimated resistance (after expansion) | | | |
|---|---|---|---|---|---|---|---|---|
| | | Pressure (bar) | Volume (liter) | Equiv. diam (mm) | FDP side (kN) | EBI side (kN) | EB toe (kN) | Total (kN) |
| C11J-P03 | 29 | 3.45 | 280 | 800 | 1300 | 550 | 710 | 2560 |

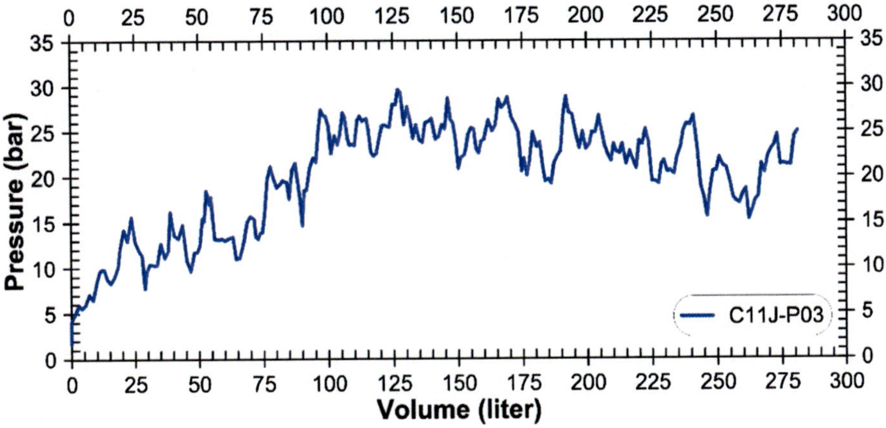


**Fig. 10.** Injection pressure-volume relationship for FDP+EBI C11-P03 (EB815)

To evaluate the design, one full-scale head-down static axial compression loading test was performed on production pile C11J-P03 using a 400 tonne hydraulic jack that applied the axial load to the center of the test pile. The loading test was performed in general accordance with ASTM D1143 (2013), Procedure B: Maintained Test to a maximum applied axial compression load of about 400 tonne (3,900 kN). The objectives of the static loading test were to confirm the axial resistance, evaluate the resulting vertical displacements, and to determine the load transfer response. The applied axial load and resulting deformations were measured using digital instrumentation (Fig. 11).

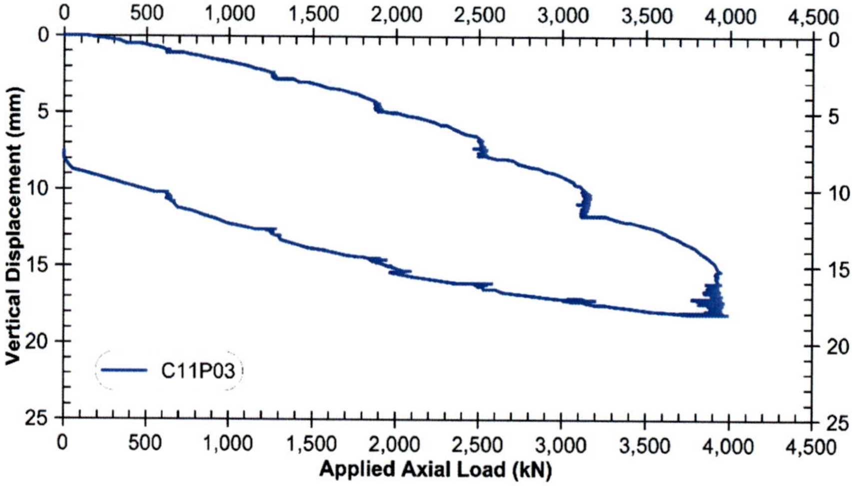


**Fig. 11.** Load-displacement response for the FDP+EB C11J-P03 test pile

The response of the FDP+EB was very stiff and experienced a vertical displacement of about 18 mm at the maximum applied test load of about 400 tonne (3,900 kN), which was about 155% of the intended unfactored working load. The permanent movement after all unloading was 8 mm (i.e., about 45% of the vertical displacement measured during the testing). At the design load of 255 tonne (2,500 kN), the resulting vertical displacement was only about 8 mm. The project required that the pile head movement at the working load be at most 10 mm, which was achieved at an applied load of 292 tonne (2,865 kN).

## 5.4 Production Piles

A summary of the measurements of the maximum injection pressures and the injected volume of grout along with the inferred equivalent diameter of the expanded EBs is provided in Table 4. A summary of inferred nominal axial resistances for the FDP+EBs is provided in Table 5. Notable variations were observed in the achieved injection pressures during grouting, which we consider due to the following:

1. The highly heterogenous soils in the Santa Cruz de la Sierra region are alluvial in origin, which means that the soil will have variable contents of fines, resulting in differences in EB injection pressure, even within a short distance between piles.
2. The sequence of grout injection, especially for large pile groups, affects the injection pressure of subsequent EBs. An EB during its expansion will densify (to varying degree) the soil out to 2 to 3 diameters from the center of the EB. Therefore, previously expanded EBs will have densified a certain zone within the group, which then will require higher injection pressures for expansion.

**Table 4.** Results from grout injection for the FDP+EBs

| Location | No. piles (ea) | Depth (m) | Injection pressure | | Injected volume | | Equivalent diameter | |
|---|---|---|---|---|---|---|---|---|
| | | | Range (bar) | Average (bar) | Range (L) | Average (L) | Range (mm) | Average (mm) |
| Overall | 334 | 29.0 | 18.2 to 75.1 | 40.9 | 123 to 286 | 275 | 411 to 795 | 770 |

**Table 5.** Estimated resistance of FDP+EBs (after expansion)

| Location | No. piles (ea) | FDP side (only) | EB side (only) | | EB toe (only) | | Total axial | |
|---|---|---|---|---|---|---|---|---|
| | | Estimated (kN) | Range (kN) | Average (kN) | Range (kN) | Average (kN) | Range (kN) | Average (kN) |
| Overall | 334 | 1600 | 420 to 1720 | 910 | 370 to 2400 | 1250 | 2500 to 5800 | 3800 |

## 6 Conclusions

This paper describes the principles, design, installation, and monitoring of Expander Bodies (EBs) combined with full displacement piles (FDPs) piles to increase its axial resistance for the support of the new World Trade Center development in Santa Cruz, Bolivia. In addition to improving the soil surrounding the device and to increasing the capacity and stiffness of the piling system, the EB also serves as a geotechnical investigation device (i.e., can be used to infer geotechnical parameters) for each of the piles onto which it is incorporated and provides direct monitoring and quality control/assurance, thereby increasing reliability and reducing the risk and uncertainty of performance. One full-scale, head-down axial compression test was performed on a production FDP+EB pile to confirm adequate resistance and behavior under a design load per individual pile of 150 tonne (1,500 kN). The maximum applied axial compression load was about 400 tonne (3,900 kN), which resulted in a vertical displacement of about 18 mm at the pile head. The measured injection pressures and volume of injected grout varied widely for the production FDP+EBs, even though the piles were tipped in the dense sand layer. The injection pressures ranged from about 1,820 to 7,510 kPa, respectively, resulting in an estimated axial resistance ranging from about 2,500 to 5,800 kN. The test results showed that equipping an FDP pile with an EB significantly improved the bearing and enabled the FDP+EB pile to be considerably shortened—from 39 m to 29 m. The shortening saved the project about 25% in pile construction costs and about 40% in construction time.

**Acknowledgements.** The authors would like to express their sincere gratitude to Dr. Bengt H. Fellenius and Dr. K. Rainer Massarsch for their comprehensive review and insightful comments and recommendations on the manuscript.

## References


ASTM D1143/D1143M-07: Standard Test Methods for Deep Foundations Under Static Axial Compressive Load. ASTM International, West Conshohocken, PA (2013). www.astm.org

Brettmann, T., NeSmith, W.M.: Advances in auger pressure grouted piles: design, construction and testing. In: Advances in Designing and Testing Deep Foundations. Geotechnical Special Publication (GSP) No. 129. pp. 262–274. ASC (2005)

Brown, D.A., Dapp, S.D., Thompson, W.R., Lazarte, C.A.: Geotechnical Engineering Circular No. 8 - Design and Construction of Continuous Flight Auger (CFA) Piles. FHWA-HIF-07-03. Federal Highway Administration, U.S. Department of Transportation, Washington D.C., 297 p. (2007)

Brown, D.A., Turner, J.P., Castelli, R.J., Loehr, J.E.: Drilled Shafts: Construction Procedures and Design Methods. Geotechnical Engineering Circular No. 10. Report No. FHWA-NHI-18-024. Federal Highway Administration. March. 758 pp. (2018)

Clarke, B.G.: The Pressuremeter in Geotechnical Design, p. 383. Blackie Academic & Professional, Cambridge (1995)

Fellenius, B.H.: Basics of Foundation Design (Red Book). Electronic. 529 p. (2018). www.fellenius.net

Fellenius, B.H., Terceros Herrera, M.: Response to load for four different types of piles. In: Proceedings of the DFI-Europe/EFFC International Conference on Piling and Deep Foundations, Stockholm (2014)

Fellenius, B.H., Terceros Herrera, M., Terceros Arce, M., Massarsch, K.R., Mandolini, A.: Static response of a group of 13 piles tested simultaneously. In: Proceedings of the 4th Bolivian International Conference on Deep Foundations (2019)

Fleming, W.G.K., Thorburn, S.: Recent piling advances, state of the art report. In: Proceedings of the International Conference on Advances in Piling and Ground Treatment for Foundations. Institution of Civil Engineers (ICE), London, pp 1–16 (1983)

Goudreault, P.A., Fellenius, B.H.: UniSettle, ver. 4, Tutorial with Background and Analysis Examples. UniSoft Geotechnical Solutions Ltd., 85 p. (2011). www.UniSoftGS.com

Goudreault, P.A., Fellenius, B.H.: UniPile, ver. 5, User and Examples Manual. UniSoft Geotechnical Solutions Ltd., 120 p. (2014). www.UniSoftGS.com

Incotec, S.A.: Personal communications (2020)

Incotec, S.A: Informe de Investigación Geotécnica (Supplemental geotechnical report), 71 p. (2017)

Marinucci, A., Terceros Herrera, M.A., Terceros Arce, M.: Quality control and estimation of axial resistance by using an expander body, a large-scale pressuremeter. In: Proceedings of the DFI 45th Annual Conference on Deep Foundations. National Harbor, MD (2020)

Marinucci, A., Terceros Herrera, M.A., Terceros Arce, M.: Increased resistance of small-diameter piles and ground anchors by using an expander body. In: Proceedings of the International Foundations Congress and Equipment Exposition (IFCEE). Dallas, TX (2021)

NeSmith, W.M.: Static capacity analysis of augured, pressure-injected displacement piles. In: Proceedings of the International Deep Foundations Congress 2002. Geotechnical Special Publication (GSP) No. 116, vol. 2, pp. 1174–1186. ASCE (2002)

Sabatini, P.J., Tanyu, B., Armour, T., Groneck, P., Keeley, J.: Micropile Design and Construction. FHWA-NHI-05-039. Federal Highway Administration, U.S. Department of Transportation, Washington,. D.C., 455 p. (2005)

Sellgren, E., Hellman, L., Skogberg, B.: Expander piling – a careful foundation and underpinning method. In: Proceedings of the 12th International Conference on Soil Mechanics and Foundation Engineering, vol. 12, pp. 1027–1030 (1989)

SiteCal Ingenieria y Arquitectura. Informe de Investigación Geotécnica – Perfil Geotécnico, 45 p. (2015)
Terceros Herrera, M., Wetterling, S., Massarsch, K.R.: Application of the Soilex pile system with expander body in Bolivia. In: X. Congreso Panamericano de Mecanica de Suelos e Ingenieria de Cimentaciones, México, pp. 1319–1327 (1995)

Check for updates

# Sustainable Civil Infrastructures Utilization and Regulations of Innovative Technology to Improve Road Safety via Drivers' Warnings and Enforcement

Lama Abufares[1](✉) and Faisal Awadallah[2](✉)

[1] University of Illinois, Urbana-Champaign, USA
lamaha2@illinois.edu
[2] Birzeit University, Birzeit, Palestine
fia@birzeit.edu

**Abstract.** The research addresses theoretical analyses for accidents' frequencies and severities in terms of speed and driver, vehicle and roadway conditions. Variable speed limits (VSL) for various road conditions, such as wet and dry pavement, visibility, day/night driving and general vehicles vs heavy vehicles are studied. Differential speed limits are theoretically analyzed via equation derivations and examples of used equations, especially by AASHTO. The paper addresses regulations and technology innovations in setting VSL for various conditions and vehicle types and how to relay VSL to drivers. In addition to warn speeding drivers and ultimately to use high-tech methods to enforce the speed limit.

The paper addresses the issue of privacy of technology sensors and trackers of drivers. It concludes with innovative technology adaptations and regulations, especially addressing legal issues. This includes vehicle 'black box' data via usage of global positioning systems (GPS) and geographic information system (GIS) up to the use of autonomous vehicles. The voluntary or mandatory usage of data by other than the driver, such as the vehicle insurance agency, the motor vehicle administration or even the police department are discussed and explored for practical uses that need to gain public acceptance when they become convinced that their benefits clearly outweigh their disadvantages. These are complex issues not just in terms of technological innovations and their validation, but in terms of regulations, legal aspects, privacy, and public acceptance. Hence, optional usage with incentives from insurance companies, motor vehicle administration and even the local police is a logical path for consideration.

**Keywords:** Roadway safety · Variable speed limits · Drivers automated warnings and enforcement

## 1 Introduction

The annual fatalities of traffic accidents worldwide have reached alarming rates. The World Health Organization (WHO) estimated traffic accident fatalities in the world in


H. Khabbaz et al. (Eds.): GeoChina 2021, SUCI, pp. 110–126, 2021.
https://doi.org/10.1007/978-3-030-79650-1_9

2016 to be 1.35 million [1]. Traffic fatalities are the leading cause for global injury mortality [2], and in 2015 motor vehicle traffic crashes were the leading cause of unintentional injury death among males for ages 5 to 23. While among females, motor vehicle traffic crashes were the leading cause of unintentional injury death for ages 3 to 23 [3]. Traumatic injuries from road traffic accidents are the third most common cause of disabilities worldwide and the second most common cause in developing nations [4]. Traffic accidents also have a high economic toll and societal impact, where the cost of motor vehicle crashes in the United States is estimated at $871 billion in 2010 & $1 trillion in loss of productivity and loss of life [5]. There are also many unquantifiable aspects associated with traffic accidents such as pain and suffering of the injured, the agony for the loss of a family member, the loss of creative minds; in addition to social and emotional detrimental effects. Therefore, traffic safety is, and should be, the utmost priority for traffic engineers. There have been extensive research and engineering practices to improve vehicle, roadway, and traffic control devices design and operation. Vehicle safety improvements in the past decades had numerous and significant impacts on reducing accidents frequency and severity such as seatbelts, airbags, anti-lock braking systems, and warning peepers when vehicles are in the blind spots of a moving vehicle. The roadway design improvements also had significant safety features in the past decades such as New Jersey median barriers for multi-lane highways, shoulders' rumble strip, and skid resistant pavement. In addition, traffic control devices had various developments to improve safety margins for signal operation and introduction of electronic speed and red light running enforcement. However, the overwhelming cause of accidents is still human error rather than environmental or vehicle-related. The National Motor Vehicle Crash Causation Survey conducted from 2005 to 2007 showed the critical reason for pre-crash was assigned to drivers in 94% of the crashes in the USA. The main driver-related reasons were recognition errors (41%) which are mostly caused by distractions, decision errors (33%) such as speeding and performance errors (11%) mostly caused by fatigue or sleepiness [6].

## 2 Background and Theoretical Analysis

The latest traffic safety data in the USA show that traffic accidents fatalities have been relatively constant in the past few years. Statistical data from the U.S. Department of Transportation – National Highway Traffic Safety Administration (NHTSA) released in September 2019 shows that the number of traffic accidents fatalities in 2016 and 2017 were 37,806 and 37,133 respectively [7]. Whereas the fatality rate per 100 million vehicle-mile traveled (VMT) in 2016 and 2017 were 1.19 and 1.16 respectively [7]. The critical pre-crash reason was attributed to drivers in 94% of the crashes for drivers at all ages as opposed to vehicle (2%) or environmental factors (2%) [8].

### 2.1 Theoretical Analysis

The basic question that is investigated and analyzed is the probability and severity of accidents for various occurring conditions. Some of the commonly occurring accident conditions that involve human factors are speeding at various conditions and accidents at intersection. These conditions are addressed in this research, particularly in regard to the basic rule of highway geometric design, Eq. (1) below:

$$\text{Available Sight Distance (ASD)} \geq \text{Stopping Sight Distance (SSD)} \quad (1)$$

Where on all locations on a road/highway the ASD should be greater than SSD. AASHTO 2018 Green Book [9] defines the available sight distance based on a driver's eye height within a design vehicle of 108 cm and an obstruction height of 60 cm and the SSD is based on the following equation:

$$SSD = 0.278vt + \frac{v^2}{254[(\frac{a}{9.81}) \pm G]} \quad (2)$$

Where SSD is the stopping sight distance (m), v is the design speed (km/h), t is reaction time in seconds with a default value 2.5 s for geometric design, 'a' is the deceleration rate (m/s$^2$), with a default value 3.4 m/s$^2$ for geometric design, and G is grade in decimal form or the percent grade divided by 100. However, the SSD in the AASHTO 1994 Green Book [10] is based on the following equation:

$$SSD = vt + \frac{v^2}{2g(f \pm G)} \quad (3)$$

Where SSD, t, and G were defined earlier, v design speed (m/s), f is the coefficient of skid resistance, and g the gravitational constant of 9.81 m/s$^2$. Equation (3) could also use any compatible units. The 1994 Green Book uses several versions of the above formula as empirical formulas with set units. The first term, the reaction distance, is the same as in Eq. 2 and only uses different units. However, the second term, the braking distance, is conceptually different; in the 2018 Green Book [9], it is based on the deceleration rate, while in the 1994 Green Book [10] it is based on the pavement forward coefficient of friction. The main contributory factors in Run-off Road fatal and injury crashes are: wet pavements, speeding, and distraction [10].

The following are analyses of specific human factors and conditions mostly in relation with the above equations, namely, alcohol-impaired driving, dry pavement versus wet pavement conditions, nighttime driving, heavy vehicles, intersections locations, and speeding:

## 2.2 Dry vs. Wet Pavement Conditions

The SSD for geometric design is based on Eq. (2) with a default comfortable deceleration rate of 3.4 m/s$^2$. However, all vehicles have braking mechanisms that could have much higher deceleration rates for emergency hard stops at dry pavement conditions. Vehicle braking mechanisms cannot exceed the maximum friction force between the tires and pavement. The friction force is dependent on the pavement type and condition, especially wet or dry pavement conditions and tires' type and condition. The SSD with tire-pavement friction limitation is stated in Eq. (3) and it was used in the AASHTO 1994 version [10] and earlier versions. There is a high correlation between Eq. (2) with a default deceleration rate of 3.4 m/s$^2$ and Eq. (3) with wet pavement coefficients as can be seen in columns 3 and 4 in Table 1 for a level roadway condition. It should be also noted that at high speeds such as 120 km/h, the SSD for wet pavement based on surface friction is sufficiently higher than when compared to the deceleration rate equation (Eq. 2). In other words, there is additional safety factor at high speed for AASHTO 2018 version, namely there is still sufficient friction to increase the deceleration rate above 3.4 m/s$^2$. The coefficient of skid resistance for wet pavement was used in AASHTO 1994 [11] conservatively with the assumption of poor tire treads and some pavement polishing. What is particularly noteworthy in Table 1 are the last two columns. The column before last (column 6) is for SSD for dry pavement and using dry skid coefficients (column 5) from the same source that provided the skid coefficients for the wet pavement, AASHTO 1994 [11]. Hence, it is evident that the SSD for dry pavement is substantially lower than that for wet pavement, especially for high speeds.

The safe speeds for dry conditions (last column, Table 1) were calculated by setting the SSD in Eq. (3) for wet pavement conditions (column 3), using the dry pavement coefficient of friction (column 4) and maintaining the default values in Eq. (3). Then solving the quadratic equations for the value of $v$, the safe speed for dry pavement is obtained. The rational values of the results of the quadratic equations are provided in the last column. Table 1 shows the safe speeds for dry pavements that were calculated from the quadratic equations are slightly higher than the design speed for wet pavement (column 1) for low design speeds, and substantially higher than the design speeds for high design speed values. Thus, for a design speed of 120 km/h, there is more than a 40 km/h increase in safe speed for dry pavement conditions, or if it would be safe to drive at a speed of 120 km/h at wet pavement conditions, it would be safe to drive at a speed of 160 km/h for dry conditions. This indeed is a substantial difference, which strongly suggests considering differential speeds during dry and wet pavement conditions.

### 2.3 Heavy Vehicles vs. General Traffic Conditions

Braking distances for heavy vehicles (trucks) are substantially longer than those for passenger cars on dry pavement. The braking distance depends on the braking force on the tires and the maximum friction force between the tires and pavement surface. The braking force depends mainly on the efficiency of the braking mechanism, the initial speed, and the vehicle's weight. For passenger cars, the braking mechanism is capable of providing substantial deceleration rates in the range of 0.9 g compared to about 0.5 g for heavy trucks [12]. Thus, using the braking distance term of Eq. 2 (the second term) Table 2 was prepared for various deceleration rates. It shows that passenger cars have braking distances about 60% of that of heavy vehicles at dry pavement, while passenger cars' dry pavement braking distances are about 40% of that on wet pavements.

The US National Highway Traffic Safety Administration standards require that the braking distance for a loaded single unit truck from an initial speed of 60 mph (97 km/h) does not exceed 310 ft. (94.5 m) for dry good pavement conditions with good tire treads with a coefficient of friction of 0.9 [13]. This corresponds to a deceleration rate of only about 0.4 g. However, for passenger cars under the same condition, the braking distance of Eq. (3) controls and equals 41 m, which is less than 50% of the braking distance for heavy trucks on dry pavement. The substantially higher braking distance of trucks compared to passenger cars and the limited maneuverability of trucks increase the

**Table 1.** Summary calculation for wet and dry stopping sight distance and safe speeds on level roads

| Design speed (km/hr) | $f_w$ (a) | Stopping sight distance (SSD) for wet pavement – AASHTO 1994 (m) | Stopping Sight Distance (SSD) by AASHTO 2018 (m) | $f_d$ (b) | Stopping sight distance (SSD) for dry pavement – AASHTO 1994 (m) | Safe speed for dry pavement conditions (km/h) |
|---|---|---|---|---|---|---|
| 30 | 0.39 | 29.9 | 31.2 | 0.66 | 26.2 | 33.2 |
| 40 | 0.38 | 44.3 | 46.2 | 0.65 | 37.5 | 45.7 |
| 50 | 0.34 | 63.6 | 63.5 | 0.64 | 50.1 | 59.3 |
| 60 | 0.32 | 85.9 | 83.0 | 0.64 | 63.8 | 73.7 |
| 70 | 0.30 | 112.8 | 104.9 | 0.63 | 79.2 | 88.7 |
| 80 | 0.29 | 142.3 | 129.0 | 0.63 | 95.5 | 103.8 |
| 90 | 0.28 | 176.3 | 155.5 | 0.62 | 113.9 | 117.2 |
| 100 | 0.28 | 209.9 | 184.2 | 0.62 | 132.9 | 133.2 |
| 110 | 0.28 | 246.3 | 215.3 | 0.61 | 154.4 | 148.9 |
| 120 | 0.27 | 293.1 | 248.6 | 0.61 | 176.2 | 163.3 |

(a) Coefficient of wet pavement skid friction values were approximated from curve 9, wet plant mixes in Fig. III-1A of the AASHTO 1994 [9]
(b) Coefficients of dry pavement skid friction values approximated from curve 8 for dry plant mixes in Fig. III-1A of the AASHTO 1994 [9]

probability of accidents. Furthermore, the greater the weight of a truck, the greater the severity of an accident, which is related to the dissipation of kinetic energy. Opponents of lower truck speed limits argue that truck drivers are more experienced, and have a higher viewing position than drivers' of passenger cars, thus they are more vigilant. Hence, the higher braking distances are partially compensated by a lower average reaction distance for truck drivers compared to passenger car drivers. The reaction distance for a 60 mph (97 km/h) speed is 67 m, based on a reaction time of 2.5 s recommended by the 2018 Green Book [9] for geometric design. Thus, if we assume that truck drivers are on the average twice as vigilant as passenger car drivers are and their average reaction time is 1.25 s, then there would be a reduction of reaction distance by 50% (33.5 m). This still leaves a difference at least 20.0 m longer SSD on dry pavements for trucks comparing to passenger cars. This significant factor should be considered as contributor to higher accident frequency among heavy vehicles.

The severity of a crash is related to dissipation of kinetic energy ($KE = 0.5\ mv^2$, where m is the mass of a vehicle and v is its velocity) [14]. Thus, accident severity is proportional to vehicle weight (comparisons with mass or weight are identical, since weight equal mass times the gravitational constant). The U.S. federal commercial vehicle maximum gross weight on the interstate highway system is 80000 lb (36287 kg) [15], while many passenger cars have weights less than 4000 lb (1814 kg). Hence, the weight

**Table 2.** Braking distances for various deceleration rates

| Design speed (km/hr) | Braking distance by AASHTO 2018 for default deceleration rate of 3.4 $m/s^2$ (m) | Braking distance by AASHTO 2018 for passenger cars on dry pavement, deceleration rate 8.5 $m/s^2$ (m) | Braking distance by AASHTO 2018 for heavy trucks on dry pavement, deceleration rate 5.0 $m/s^2$ (m) |
|---|---|---|---|
| 30 | 10.2 | 4.1 | 7.0 |
| 40 | 18.2 | 7.3 | 12.4 |
| 50 | 28.4 | 11.4 | 19.3 |
| 60 | 40.9 | 16.4 | 27.8 |
| 70 | 55.7 | 22.3 | 37.8 |
| 80 | 72.7 | 29.1 | 49.4 |
| 90 | 92.0 | 36.8 | 62.6 |
| 100 | 113.6 | 45.4 | 77.2 |
| 110 | 137.4 | 55.0 | 93.5 |
| 120 | 163.6 | 65.4 | 111.2 |

of heavy trucks could be 20 times heavier than passenger cars making trucks' accidents much more severe.

There have been many regulations worldwide to reduce heavy vehicle speed compared to general traffic including differential speed limits. In May 2015, there were only five states that have differential speed limits for Trucks, namely California, Idaho, Michigan, Montana, and Washington. The speed limit difference between trucks and general traffic in these states range between 10 and 15 mph [16]. Speed limiter of 90 km/h have been required to be installed on Trucks (12 tons and more) entering the European Union since 1994. In addition, Australia and Canada have introduced speed limiter and the USA is in an advanced stage for consideration and legislation of speed limiters [17, 18].

## 2.4 Daytime vs. Nighttime Safety Conditions

Available sight distance (ASD) during daytime and good visibility conditions are many times greater than the SSD for most roads, especially rural expressway segments. Only in certain horizontal and vertical curves, the ASD is close to the SSD. Thus, it is only for the most critical locations of horizontal and vertical curves that drivers have only 2.5 s unexpected reaction time before they apply the brakes and stop in time to avoid an accident. However, at night, the ASD provided by a vehicle's headlights has a fixed coverage area regardless of vehicle's speed. The ASD or visibility distance at nighttime is very complex and varies significantly in various studies, depending on numerous factors. Certainly, the ASD provided by a vehicle's headlights is confined mainly to the direction straight ahead of a vehicle. However, even on straight level sections of roadway, the ASD provided by a vehicle's headlights varies greatly, subject to the retro-reflectivity, contrast and general physical characteristics of an object; in addition to the intensity of the headlights (especially low beam and high beam), and their alignment and position on a vehicle. Furthermore, the driver's visual characteristics and glare from oncoming headlights are also factors that affect the ASD [12]. Most research on nighttime visibility distance has been conducted on signs, markings and pedestrians. Awadallah [19] showed there is serious limitations of the retro-reflectivity for warning signs providing sufficient visibility distances at high speeds. Pedestrian recognition distances of 50 m or less for black and grey pedestrians' clothing have been reported [20, 21]. The average recognition distance for pedestrians in no-glare conditions was 76.5 m, with distances being shortest for darkly clad pedestrians (13 m). Headlight beam also had a significant effect on recognition distance; overall, changing from low beam to high beam improved recognition distances from a mean of 59 m to 94 m, with recognition distances improving by a factor of 3.5 times for black clad pedestrians [22]. Thus, nighttime recognition distances are very sensitive to numerous variables and must be interpreted carefully. The speed limit should ensure that ASD for nighttime driving on unlit highways is greater than SSD, especially for wet pavement conditions. Thus, the above research results suggest that speed limits lower than 100 km/h are needed for recognition distances of objects during low-beam lighting in order for the ASD to be equal to or greater than the SSD. For low reflectivity and contrast objects, the safe speed at night is much less than 100 km/hr. Animals crossing highways at night are more likely to cause an accident than during daytime, because of the limited area lit by a vehicle's headlight, especially on the sides of the road.

### 2.5 Intersections Safety

Red light running is a serious intersection safety issue. Intersection areas represent a very small percentage on the entire roads system. However, about 40% of motor vehicle crashes in the US in 2008 occurred at intersections [23]; in addition, intersections are a major source of traffic delays and roadway capacity constraints; because intersections have concentration of conflict points in a limited area. There were more than 2.3 million reported intersection-related crashes, resulting in more than 7,770 fatalities and approximately 733,000 injury crashes in 2008 in the USA [24]. Intersections have a high number of traffic flow conflicts and thus many practical solutions, particularly temporal and spatial separations have been used. In addition, many researches have addressed intersection safety and delay. Number of researchers have addressed minimization of intersection delay and improving safety at these conditions through green extension and advance warning flashers [23–25]. However, considerably fewer researches addressed intersection regulations issues [26, 27]. Thus, since more than 90 percent of the causes of accidents is due to human factors, there is an essential need to address the issues of regulations, awareness, and enforcement at intersections that would limit the number and severity of accidents.

### 2.6 Speeding Accidents

Speeding is one of the most dominant factors contributing to traffic crashes. NHTSA estimates that the annual economic cost to society of speeding-related crashes is $40.4

billion [28]. In 2012, speeding was a contributing factor in 30 percent of all fatal crashes. Speeding-related fatalities increased by 2 percent from in 2012. In 2017, 9,717 people were killed due to speeding, which forms about 26% of total traffic fatalities [29]. Surprisingly speeding, tailgating and other risky driving crashes were found strongly related to familiar drivers which are more prone to distractions rather than unfamiliar drivers [30].

Speed was found proportionally related with both crash frequency and severity and that this relationship is described better by a quadratic rather than a linear relationship [31]. Speed is estimated to have been a factor in over a third of traffic accidents deaths in Australia and New Zealand [32]. Implementation of speed control and enforcement, especially through speed camera significantly contributes to reduction of accident frequency and severity [33, 34]. Furthermore, as indicated earlier the severity of crash is related to dissipation of kinetic energy, 0.5 $mv^2$ [13]. Hence, speed affects crash severity exponentially.

# 3 Recommendations

It is of principal importance that for driving rules and regulations to be effective, they should be legislated nationally with preferable international consensus and should be clear and well known to drivers all around the world.

Based on the data analysis that clearly shows human factors as the overwhelming cause of traffic accidents; in addition, based on the literature review and theoretical analysis that outlined the main factors affecting accidents frequency and severity, recommended regulations and technology advancements for warnings and enforcements are outlined here thereafter. The recommendations do not include options in wide usage such as truck speed limiters. It is important to note that the following recommendations are intended for discussions among traffic engineers, researchers, and policy makers on a national and international levels.

## 3.1 Regulations to Improve Safety

### 3.1.1 Differential Speed Limits

The above analysis demonstrated that the safe speed limit for wet and dry conditions should vary depending on speed as per Table 1. In addition, for other weather conditions, such as snow, sleet, fog, sand storms the speed need to be reduced significantly. Table 2 shows the braking capabilities of heavy vehicles at dry pavement is less than passenger cars. For example, the braking distance for heavy trucks at 80 km/hr is 49.4 m, whereas less (better/safer) braking distance is achieved for passenger cars at 100 km/h, namely 45.4 m. As for the nighttime sight distance for unlit highways and streets and only relying of low-beam (low-beam is necessary when there is significant opposing traffic), the maximum safe speed should not exceed 100 km/h, but also this varies according to brightness and contrast of the target. Thus the posted speed limits that are set for all vehicles and conditions are not accurate. In some cases, they could be high and unsafe, while at other conditions it may be safe to exceed the speed limit by more than 40 km/h as per passenger cars driving on dry pavement of speed limit of 120 km/h (Table 1).

Differential speed limit regulations for various conditions could become complicated, especially if these differential speed limits are posted. Hence, it is recommended to use regulations to reduce speed at certain conditions. Thus the posted speed would be for the default for general traffic (excluding heavy vehicles), dry and daytime conditions. The speed limits between 70 km/h and 100 km/h are suggested to be reduced by 5 km/h for each of the three conditions (heavy vehicles, wet pavement, and nighttime); while for posted speed limits above 100 km/h, it is suggested to reduce the speed limit by 10 km/h for each of the three conditions. Noting that at any of the above conditions, the difference of speed limits between heavy vehicles and general traffic would only be one increment (5 km/h for 100 km/h or less and 10 km/h for more than 100 km/h). Furthermore, the posted speed limit (before this suggestion of differential speed limit regulation) applies for wet conditions, thus it may be increased (in the posted sign) where deemed appropriate by 5 km/h or 10 km/h or more, especially depending on the reduction requested during wet pavement, e.g., 5 km/h or 10 km/h and considering the results of Table 1. Second the reductions of 5 km/h and 10 km/h for the various conditions is generally less than needed based on the analysis of Sect. 2; however, what is more important is to stress the danger for speeding for the outlined three conditions. Thus, it is hypothesized that by statutory regulation of speed reductions at these conditions, drivers most likely would reduce speed, but more importantly they are expected to be more vigilant.

### 3.1.2 Yield Before Go at the Onset of Green

When a traffic accident occurs due to red light running, police officers easily determine the responsibility of the accident. Almost all researches, awareness campaigns and enforcement efforts are concentrated on addressing the violators of this serious traffic rule. This should not belittle. However, the emphasis for reducing intersection crashes should not be solely directed towards red-light runners; drivers at the opposing approaches, which have the onset of the green indication, should yield the right-of-way to vehicles within the intersection or fast approaching vehicles on red interval approaches. The Manual of Uniform Traffic Control Devices (MUTCD), 2009 [35] states that "Vehicular traffic facing a CIRCULAR GREEN signal indication is permitted to proceed ........such vehicular traffic, including vehicles turning right or left or making a U-turn movement, shall yield the right-of-way to (a) Pedestrians lawfully within an associated crosswalk, and (b) Other vehicles lawfully within the intersection." The word "lawfully" is recommended to be omitted and to add to 'vehicles within the intersection', vehicles imminent to enter the intersection after the onset of its red interval (speeding into the intersection). Thus, it is recommended that drivers on the onset of green should yield the right-of-way to vehicles and pedestrians already in the intersection or fast approaching the intersection in order to avoid serious and imminent accidents. However, the responsibility of an accident due to a driver entering the intersection during the red interval should not be equal to the responsibility of a driver entering the intersection at the onset of green without yielding the right of way to vehicles within the intersection. It suggests that the responsibility of a driver entering the intersection at the onset of green without yielding the right of way to vehicles within or approaching the intersection should be 10% to 30% of that of a driver entering the intersection during the red interval. Thus, a driver not yielding the right of way to vehicles within or approaching the intersection at the onset of green could be charged with up to 30% of the amount of the citation fee for red light runners. In addition, his/her insurance company could be responsible for up to 30% of the damages for an accident caused mainly by a driver entering the intersection during the red interval. Certainly, this modification of the above traffic ordinance should be lawful and binding to motorist and it should be communicated to drivers, especially through drivers' license tests and awareness campaigns.

The main intention of this regulation is to reduce severe accidents. Even though if there were no legal responsibilities to do so, drivers would certainly avoid potential crashes that could cause them injury or death.

## 3.2 Use of Advance Technology to Improve Safety

### 3.2.1 Variable Speed Limits

Posted variable speed limits existed for many decades; however, they are not practical to have them on all highways and streets. New technology advancements with Global Positioning System (GPS) and Geographic Information System (GIS) could become rather inexpensive and practical to use on all vehicles, even old vehicles could have an added equipment which would be similar to smart phones. Speed limits could be posted on each segment of highway or street (as existing speed) with variations for various conditions. Thus, wherever a vehicle is moving on a highway/road network the speed limit on such segment would be posted for a driver on a vehicle screen. This speed limit could be variable for various conditions not only for the three conditions outlined in Sect. 2, but also the speed limit could be set for various weather conditions such as snow, sleet, fog or sandstorms and other conditions such as roadway construction and traffic incidents. The official highway agencies would be in charge for making set changes based on specific conditions. For general conditions like wet pavement and nighttime visibility this could be set for each vehicle based on sensors within vehicles or for all vehicles based on sunrise and sunset times and the weather monitoring conditions at various locations. Each weather station would have a cordon line of road network that it covers. Of course, specific conditions like construction and highway incidents would be input manually by the highway agency officials.

### 3.2.2 In-Vehicle Drivers' Guidance, Warnings and Citations

The technology described in Sect. 3.1.1 above for in-vehicle VSL screens could be used as additional guidance for drivers. Drivers who do not have such technology in their vehicles must rely on the posted speed limits and related regulations to adhere to the speed limits at various highway/street segments. However, if there are speed limits for specific temporary conditions such as fog, snow, construction this only could be communicated via in-vehicle devices, this could only be legal if all vehicles have such devices. Furthermore, the issue of equipment malfunction and no signal for mobile phone/GPS is also another issue that hinders such technology not applicable at the present time. The following technology concepts may be applied on optional basis in the near future:

#### 3.2.2.1 Speeding

The technology for posting speed limit within a vehicle screen is certainly readily available; however, with advance technology and regulations additional features are certainly possible such as setting the vehicle cruise at the set speed limit. Furthermore, variable speed limit could be posted for various conditions. This system/application can also warn speeding drivers that exceeding the speed limit by certain value (e.g., 5 km/h for speed less than 80 km/h and 10 km/h for speed 80 km/h and more).

Another issue is to share the speeds of vehicles with insurance companies, motor vehicle administrations and even police departments. The owner of the vehicle could be held responsible for traffic violations or the identity of the driver could be determined via the eye iris or the finger print. This certainly is a privacy and legal issue to enforce such sharing of data. Optional sharing of define parts of this data for specific agencies ought to be considered, and certainly before any legislation for mandatory sharing of some data. Only defined data such as exceeding the speed of 10 km/h or more for speed less than 100 km/h and 15 km/h for speeds 100 km/h or more with time and location of the violation. This specific information may be provided for insurance companies on the condition that drivers with less than some certain violations per year (e.g., 3) may have certain percentage of insurance reduction (e.g., 20%). Some drivers may have to share some data with the motor vehicle administration and the police department as a condition for not revoking a license due to speeding tickets. Drivers that share data with the police department could get audio/visual warning before reaching set value for speed violation. Thus if a driver ignores the warning and exceeds the set value of exceeding the speed limit, she/he would get reported to the police department and would be issued a speeding ticket. A speeding ticket may be issued and delivered to a driver automatically and immediately via one or more methods (e.g., sms, email, and announcement through the same device used in the vehicle for this purpose, etc.).

#### 3.2.2.2 Wrong-Way Driving

Similar to principle of speeding, accurate GPS that shows the exact location of a vehicle and direction of motions from comparison of GPS coordinates vs. time (a second, two seconds, three second earlier) would provide the direction of travel of a vehicle. This information may be compared to direction of travel of roadway and one-way streets and ramps. For example, if a vehicle enters a wrong way ramp on a highway, the device would strongly alarm the driver of driving on the opposite direction and command the driver to pull to the side and turn around. Other than strong warnings and possibly driver citation, other measures such as limiting the speed of the vehicle and making the driving jerky to force a driver not to proceed in this very dangerous driving violation.

#### 3.2.2.3 Traffic Signals Red Light Crossing

Cameras posted on traffic signals are common in photographing vehicle tags inside intersections when the signal is red for a given approach. Automatic citation for red light crossing usually includes the signal location, date and time and specific approach.

This could be done for all intersections without cameras at the intersections, but with a camera fixed on top of the inside of the windshield where it is connected to the GPS/GIS special program application. The location of a vehicle relative to the stop line could be determined from the GIS/GPS applications, particularly when the vehicle is in motion. If the vehicle camera takes a photo of the signal head showing a red indication just before the vehicle crosses the stop line, then this vehicle may be subject to red light crossing citation with basically the same documentations as per a camera posted on signal citation documentations. In addition, this system could warn drivers to stop on the onset of yellow interval if the vehicle is driving before the decision line. The decision line is where a vehicle located before this line and driving at the speed limit could and should stop at the approach stop line (using default comfortable deceleration rate of 3.4 $m/s^2$) if the onset of yellow occurs before the vehicle crosses decision line, otherwise the vehicle may continue at the speed limit and cross the intersection safely (red interval starts after the vehicle passes the stop line) [36]. An approximate calculation of the decision line could be performed given the speed of approach and the speed of the vehicle, but a more precise calculation of the decision line could be performed if the length of the yellow interval is provided in the attribute of the GIS program.

#### 3.2.2.4 Tailgating

Tailgating via radar distance measurements are currently used on many vehicles. These types of sensor may be used with speed odometers to determine what is the minimum distance between vehicles at various speeds. Therefore, drivers may be warned of driving too closely and if they get dangerously close they will be warned that they will get a citation. Citations only should be possible if the system is mandatory or as precautionary measures for drivers with reckless driving records and as a condition for probation driving license.

### 3.2.3 In Vehicle Black Box and Miscellaneous Warning and Citation Conditions

There are numerous other conditions that with advance high tech monitoring drivers may be warned or given citation for traffic violations. Some of the conditions include driving under the influence, drowsiness or exceeding the hours of driving per day, and performing illegal turns. All types of violations may be summarized per month with information of the type of violation, date/time, location and specific conditions as weather, construction, etc. If all violations are preceded by warning to the drivers and if records of the violations (collected within the vehicle black box) are shared with an insurance company and police department, where citation could be issued, it is expected that substantial decrease of drivers' errors and violations will occur.

### 3.2.4 Privacy, Optional and Mandatory Monitoring

The drivers' guidance and warnings for the above outlined conditions are recommended to be optional and encouraged on new model vehicles. The information only provided to drivers without sharing of the data with anyone. Optional sharing of specific and limited information for specific agencies would be encouraged through incentives, such as providing certain driving performance information for an insurance company on the condition if the driver (drivers of the vehicle) perform better than set criteria for driving performance the vehicle insurance fees would be reduced by a set percentage. If a driver consent to share information with motor vehicles administration and police department, she/he would be liable for citations (if warnings are ignored), but there should be incentive of reduced registration fees or vehicle tax. However, some drivers could only have a driver license privilege if they consent to share such information. These cases may be for drivers with bad driving records or drivers below a certain age (e.g., vehicles used by drivers less than 18 or 21 years old). The record may only be shared if a below age driver is using a vehicle, especially using iris or thumb identification of driver's identity. Such probation drivers would be allowed to drive one or two specific vehicles during the probation periods.

In order for these types of technology to be mandatory they need to be experimented on optional basis for several years and then there should be studies for public acceptance based on various advantages and incentives. Most important is the legal aspects considering national laws and regulations including personal rights and privacy.

### 3.2.5 Autonomous Vehicles

Full automation or self-driving is now being tested in the United States, an ongoing research initiated by Smart Transportation Infrastructure Initiative (STII) is working on testing full autonomous and connected vehicles' performance in Illinois Autonomous and Connected Track (I-ACT). There exist 5 levels of automation defined by NCHRP with the first 3 levels already in use. However, "complete automation" or self-driving (level 5) is still under testing and research but a number of self-driving vehicles are now operating in Arizona (Tesla vehicles operated by Uber). This leading technology is believed to reduce total accidents; since it omits the human interference using sensors, cameras (video image processing) and programing to avoid collision. They could also choose a safe operating speed by using algorithms such as the Friction Diagram Method (FDM) which chooses a safe speed depending on visibility, speed and friction of a design critical vehicle [37, 38]. Autonomous Vehicles (AV) have many more sensors and cameras (at various positions of a vehicle) than the human sensory capabilities. Decisions are taken in less time than humans, consequently, all human related driving mistakes as driving under the influence, poor directional control, red light running, drowsy driving, speeding and others will be highly reduced which makes it safer than existing vehicles. But those safety benefits are expected to be revealed between 2040 s and 2060 s as Litman estimates [37]. Also, it would reduce congestion and delay by usage of shorter gap distance -while still ensuring safety- optimizing intersection crossing without need for signals. Information would be exchanged among vehicles and assigning the right-of-way to one while letting the others slowdown or stop, if needed. Furthermore, improve fuel efficiency and reduce air pollution and on the long run improve land use. Such technology would also offer possible job growth, independence for the elderly, children under 16 and individuals with disabilities, and even economic prosperity [38] by allowing individuals to work while traveling and reducing insurance premiums. On the other hand, the public acceptance plays a huge role in this technology's adoption and success, surveys have uncovered the public's safety concerns regarding this technology, only a quarter of surveyed people consider AVs safer than traditional vehicles and only half of them would consider using them [39]. Additionally, the cybersecurity aspect arises; internet connected systems might be hacked, Gora and Rub also stated that the connection between automated vehicles could suffer from handling a huge quantity of data too [40].

Despite the revolutionary potentials of this technology, it is not expected to be in practical usage for decades, particularly due to legal aspects, especially liability and accountability in the cases of accidents as it transfers the liability from drivers to vehicles and thus from drivers to manufacturers, who will become reluctant to manufacturing such vehicles due to this huge burden. However, a new insurance approach: "no-fault" is beginning to appear as the perfect solution for this liability problem -which is currently the law in 12 states- that allows the victims to recover their damage from their own insurers no matter whose fault was the accident.

## 4 Future Concerns

All presented regulations should be evaluated/verified through real implementation and data collection. Case studies could be performed for each technology/regulation proposed along with awareness campaigns before actual legislation to assure the accuracy of the data and thus the predictability of the results. Further research could address the social and liability complexities of the aforementioned regulations and technologies. More research efforts could be done to mathematically validate the suggested thresholds and/or modify them as needed like the suggested differential speed limits.

For future research concerning autonomous vehicles implementation, some of the suggestions would be to reinvestigate all geometric design principles and modify them to remove the driver's effect and result in more efficient road designs. A likewise investigation has been done in previous papers [41] but still no real modifications are done. Moreover, research should be conducted about mixed traffic operations; automated vehicles are expected to operate with traditional vehicles on the same roads for at least few decades of operation. Driverless vehicles will need to adapt and understand other drivers' behaviors and reactions. Further research could be done to transform the existing infrastructure into an intelligent one by providing special signs, markings and/or sensors to enable vehicle-to-infrastructure (V2I), vehicle-vehicle (V2V) and infrastructure-to-vehicle (I2V) communications.

## 5 Conclusion

There should be emphasis on reducing driver error since it accounts for more than 90% of accidents' causes by various surveys and accident reports. The research stresses considerations of safety regulations along with methods of warning and enforcement with the help of new advances in technology. The success of such approach requires addressing the technical, legal, economic and social aspects by all stakeholders.

Implementation of the provided recommendations needs extensive awareness and education campaigns for drivers and sufficient training for police officers and judicial personnel. The provided recommendations should incite focused discussions at all levels in order to refine practical and effective safety regulations, methodologies for warnings of drivers, and enforcements.

**Declarations.**

**Funding.** Self-funding.

**Conflicts of Interest/Competing Interests.** None.

**Availability of Data and Material.** Not applicable.

**Code Availability.** Not applicable.

## References

1. Global status report on road safety 2018. World Health Organization, Geneva (2018)
2. Peden, M., et al.: World report on road traffic injury prevention. World Health Organization, Geneva (2004)
3. Traffic Safety Facts. National Highway Traffic Safety Administration (NHTSA), US Department of Transportation. Washington, D.C. (DOT HS 812 499), February 2018
4. Murray, C., Lopez, A.: Alternative projections of mortality and disability by cause 1990–2020: Global Burden of Disease Study. Lancet **349**, 1498–1504 (1997)
5. The Economic and Societal Impact of Motor Vehicle Crashes, 2010. National Highway Traffic Safety Administration (NHTSA), US Department of Transportation, 2010 (DOT HS 812 013), May 2015 (revised)
6. Critical Reasons for Crashes Investigated in the National Motor Vehicle Crash Causation Survey. National Highway Traffic Safety Administration (NHTSA), US Department of Transportation. Washington, D.C. (DOT HS 812 506), March 2018
7. Traffic Safety Facts. National Highway Traffic Safety Administration (NHTSA), US Department of Transportation. Washington, D.C. (DOT HS 812 794), September 2019
8. Traffic Safety Facts. National Highway Traffic Safety Administration (NHTSA), US Department of Transportation. Washington, D.C. (DOT HS 812 630), November 2018
9. A Policy on Geometric Design of Highways and Streets, American Association of State Highway and Transportation Officials (AASHTO), Washington, D.C. (2018)
10. Intini, P., Berloco, N., Ranieri, V., Colonna, P.: Geometric and operational features of horizontal curves with specific regard to skidding proneness. Infrastructures **5**(1), 3 (2020). https://doi.org/10.3390/infrastructures5010003
11. A Policy on Geometric Design of Highways and Streets, American Association of State Highway and Transportation Officials (AASHTO), Washington, D.C. (1994)
12. Awadallah, F.: Differential speed limit analysis and regulations on rural expressways. Road Transp. Res. J. **16**(4), 16–25 (2007)
13. Electronic Code of Federal Regulations (e-CFR). Standard No.517.121: Air Brake Systems, Federal Motor Vehicle Safety Standards, National Highway Traffic Safety Administration, Department of Transportation Title 49: Transportation. http://ecfr.gpoaccess.gov. Accessed June 2015
14. Managing speed – review of current practice for setting and enforcing speed limits, Transportation Research Board (TRB) Special Report 254, Transportation Research Board, National Research Council, Washington D.C. (1998)

15. Commercial Vehicle Size and Weight Program. US DOT: Freight Management and Operations. http://ops.fhwa.dot.gov/Freight/sw/overview/index.htm. Accessed May 2015
16. Speed Limit Laws - May 2015. Governors Highway Safety Association (2015). http://www.ghsa.org/html/stateinfo/laws/speedlimit_laws.html. Accessed June 2015
17. Safety Impacts of Speed Limiter Device Installations on Commercial Trucks and Buses: A Synthesis of Safety Practice - CTBSSP Synthesis 16. Transportation Research Board of the National Academies, Washington, D.C. (2009). http://onlinepubs.trb.org/onlinepubs/ctbssp/ctbssp_syn_16.pdf. Accessed May 2015
18. Summary Report - Assessment of a Heavy Truck Speed Limiter Requirement in Canada. Transport Canada, Government of Canada. https://www.tc.gc.ca/eng/motorvehiclesafety/tp-tp14808-menu-370.htm. Accessed May 2015
19. Awadallah, F.: Prediction of the service life of warning signs. Public Roads **51**(4), 116–122 (1988)
20. Luoma, J., Schumann, J., Traube, E.: Effects of retroreflector positioning on nighttime recognition of pedestrians. Accid. Anal. Prev. **28**(3), 377–383 (1996)
21. Chrysler, S., Danielson, S., Kirby, V.: Age difference in visual abilities in nighttime driving field conditions. In: Proceedings of the Human Factors Ergonomics Society 40th Annual Meeting, Philadelphia, PA, pp. 923–927 (1996)
22. Wood, J., Tyrrell, R., Carberry, T.: Pedestrian visibility at night: effects of pedestrian clothing, driver age, and headlamp beam setting. In: 82nd Annual Meeting of the Transportation Research Board, TRB, Washington D.C. (2003)
23. Crash Factors in Intersection-Related Crashes: An on Scene Perspective, National Highway Traffic Safety Administration (NHTSA), U.S. Department of Transportation (USDOT). https://crashstats.nhtsa.dot.gov/Api/Public/ViewPublication/811366. Accessed November 2019
24. Traffic Safety Facts 2008. National Highway Traffic Safety Administration (NHTSA), U.S. Department of Transportation (USDOT). http://www-nrd.nhtsa.dot.gov/Pubs/811170.PDF. Accessed May 2015
25. McCoy, P., Pesti, G.: Improving dilemma zone-protection of advance detection with advanced-warning flashers. Transp. Res. Rec. J. Transp. Res. Board 11–17 (2003). 171844, Transportation Research Board of the National Academies, Washington, D.C.
26. Gibby, A., Washington, S., Ferrara, T.: Evaluation of high-speed isolated signalized intersections in California. Transp. Res. Rec. J. Transp. Res. Board 45–56 (1992). No. 1376, Transportation Research Board of the National Academies, Washington, D.C.
27. Sayed, T., Vahidi, H., Rodriguez, F.: Advance warning flashers: do they improve safety? Transp. Res. Rec. J. Transp. Res. Board 30–38 (1999). No. 1692, Transportation Research Board of the National Academies, Washington, D.C.
28. Traffic Safety Facts 2012 Data: Speeding. National Center for Statistics and Analysis. National Highway Traffic Safety Administration (NHTSA), U.S. Department of Transportation (USDOT), May 2014. http://www-nrd.nhtsa.dot.gov/Pubs/812021.pdf. Accessed May 2015
29. Traffic Safety Facts. National Highway Traffic Safety Administration (NHTSA), US Department of Transportation. Washington, D.C. (DOT HS 812 687), May 2019
30. Intini, P., Colonna, P., Berloco, N., Renieri, V., Ryeng, E. The relationship between familiarity and road accidents: Some case studies. Transport Infrastructure and Systems, London, UK (2017)
31. Imprialou, M., Quddus, M., Pitfield, D.: Exploring the role of speed in highway crashes: pre-crash-condition-based multivariate bayesian modeling. In: TRB 94th Annual Meeting Compendium of Papers. Transportation Research Board, Washington D.C., 11–15 January 2015

32. van Lamoen, N.: The impacts of a reduced speed enforcement tolerance threshold on road safety outcomes. In: Proceedings of the Australasian Road Safety Research Policing Education Conference, Melbourne, Australia (2014)
33. Izadpanah, P., Thukral, S., Zarei, H., Hadayeghi, A.: Safety evaluation of red light camera and intersection speed camera programs in Alberta. In: TRB 94th Annual Meeting Compendium of Papers. Transportation Research Board, Washington D.C., 11–15 January 2015
34. Høye, A.: Speed cameras, section control, and kangaroo jumps–a meta-analysis. Accid. Anal. Prev. **73**, 200–208 (2014)
35. MUTCD 2009 Edition with Revision Numbers 1 and 2 incorporated, dated May 2012 (PDF) US DOT, FHWA, p. 450. http://mutcd.fhwa.dot.gov/pdfs/2009r1r2/pdf_index.htm. Accessed June 2015
36. Awadallah, F.: Yellow and all-red intervals: how to improve safety and reduce delay? Int. J. Traff. Transp. Eng. (IJTTE) J **3**(2), 159–172 (2013)
37. Litman, T.: Autonomous Vehicle Implementation Predictions. Implications for Transport Planning. Victoria Transport Policy Institute, Victoria, Canada (2017). http://www.vtpi.org/avip.pdf
38. Jankausky, E.: Investment for Growth. Retrieved from Illinois Center for Transportation (ICT), 1 November 2019. https://ict.illinois.edu/2019/10/30/university-of-illinois-4m-investment-keeps-autonomous-vehicles-on-track/. Accessed Dec 2019
39. Piao, J., McDonald, M., Hounsell, N., Graindorge, M., Graindorge, T., Malhene, N.: Public views towards implementation of automated vehicles in urban areas. Transp. Res. Procedia **14**, 2168–2177 (2016)
40. Gora, P., Rüb, I.: Traffic models for self-driving connected cars. Transp. Res. Procedia **14**, 2207–2216 (2016)
41. Intini, P., Colonna, P., Berloco, N., Renieri, V.: Rethinking the main road design concepts for future Automated Vehicles Native Roads. European Transport. Polytechnic University of Bari, Italy (2019)

# Author Index





H. Khabbaz et al. (Eds.): GeoChina 2021, SUCI, p. 127, 2021.
https://doi.org/10.1007/978-3-030-79650-1